Water Communication

Water Communication

Analysis of Strategies and Campaigns from the Water Sector

Dr. Céline Hervé-Bazin

IWA PUBLISHING

Published by **IWA Publishing**
Alliance House
12 Caxton Street
London SW1H 0QS, UK
Telephone: +44 (0)20 7654 5500
Fax: +44 (0)20 7654 5555
Email: publications@iwap.co.uk
Web: www.iwapublishing.com

First published 2014

British Library Cataloguing in Publication Data
A CIP catalogue record for this book is available from the British Library

ISBN: 9781780405216 (Paperback)
ISBN: 9781780405223 (eBook)

Contents

About the authors

Lead Author

Dr. Céline Hervé-Bazin (France)

Since 2005, Céline has specialised in water from a communication science approach. She conducts research projects on water and women; scientific mediation, EU funded projects and knowledge and citizens and water stakeholders' perceptions. She works as a consultant in communication and has conducted several campaigns and projects for organisations from the water sector in print format as well as via TV production and Internet. She teaches communication at the CELSA, School of Communication at the University of Sorbonne.
http://celinehervebazin.wordpress.com
Contact: celinehervebazin@gmail.com

Contributors (alphabetical order)

Patricia Bakir (Jordan/United States of America)

Patricia Hotchkiss Bakir is an independent consultant specializing in institutional and communication aspects of water/environmental sector development and management. She has conducted institutional assessments and developed communication and stakeholder consultation plans on a range of projects including water utility management, water reuse, water privatization and participatory irrigation management. She is a member of the Leadership Team of the International Water Association Public and Customer Communication Specialist Group.

Alex Bielak (Canada)

Dr. Alex Bielak is an internationally-recognized, award-winning Knowledge Translation and Brokering authority and has delivered a variety of successful communications campaigns and broadly-adopted initiatives to link science with policy development. He has extensive experience in the water sector: Dr. Bielak was Senior Research Fellow and Knowledge Broker with the United Nations University Institute for Water, Environment and Health, and served as Senior Advisor to the Chair of UN-Water. He worked at senior levels in various NGOs and Canadian provincial and federal departments, and at the National Water Research Institute: Alex remains affiliated with the two leading water research groups in Canada (member of the Canadian Water Network Watershed Research Consortium Management Committee, and expert reviewer on Knowledge Application, and Honorary Member of the Canadian Rivers Institute). An Adjunct Professor in the Department of Communication Studies and Multi-Media at McMaster University, he recently founded his own company, *Alex Bielak Communications*. While he likes a nice cool glass of water, he sometimes prefers a good Pinot. (www.linkedin.com/in/alexbielak)

Kajsa E. Dalrymple (USA)

Kajsa E. Dalrymple is an Assistant Professor in the School of Journalism and Mass Communication and a faculty associate of the Water Sustainability Initiative at the University of Iowa. Her research focuses on the intersection between science, communication, and public policy. She has worked closely with both state and governmental officials and Extension efforts to help facilitate campaign development dealing with environmental conservation and assess the behavioral impact of prosocial marketing campaigns.

Sara Dolnicar (Australia)

Sara Dolnicar is a Research Professor in the Faculty of Business, Economics and Law at the University of Queensland. One of Sara's key research areas is the assessment and improvement of measures in the social sciences. She has primarily applied her work to tourism and social marketing (public acceptance of water from alternative sources, environmental volunteering and foster care). Since 2008 Sara has – jointly with Dr. Anna Hurlimann from the University of Melbourne – worked on a Australian Research Council funded project focusing specifically on beliefs the general public holds about water from alternative sources as well as water conservation, what the public's acceptance levels are of those options and how their beliefs and behaviors can be modified to be more open minded towards methods of helping countries secure a guaranteed water supply for the future.

Brita Forssberg (Sweden)

Manager of Communications at the Stockholm Water Company until 2005, Brita has been working to build the awareness and education of people in Stockholm on their active role in preserving the quality of their water and wastewater services. The results were regularly monitored and proved to be quite effective. She was the Chair of the IWA Specialist Group on Public and Customer Communications until 2012 and earlier Vice Chair of the same group until 2005. From 2006 until 2013, she was the Secretary of IWA Sverige, the Swedish National Committee. She holds a BA in languages and pedagogy at the University of Stockholm and studies of economics at the Stockholm School of Economics.

Beatriz Medina (Spain)

Beatriz Medina is a specialist in Environmental Communication based in Barcelona with AMPHOS21, an international consultancy. She has worked on numerous projects in communication and environmental science including Water Research to Market (LIFE+) and WaterDiss2.0 (FP7). She is experienced in developing methodologies to improve the effectiveness of the Science/Policy interface (knowledge brokerage) in the water sector. Her areas of expertise include analysis of environmental policy-making, public perception and design of methodologies in communication and general public involvement. She also has experience in water management and resource efficiency.

Fredrick Mugira (Uganda)

Fredrick Mugira is a multiple award-winning journalist with over 13 years of wide-ranging experience working in Uganda, pan-African and international media, and over five years' experience in training of journalists. His journalism career has concentrated on investigating and reporting stories on rural WASH issues. He has both a Master's Degree and Post-Masters Certificate in Communication for Development. He heads Water Journalists Africa - the network of journalists in Africa who report on water and sanitation. He is also the ambassador for the Netherlands based The Water Channel.

Isaac Nyengue Bahanak (Cameroun)

Isaac graduated from the National School of Agro-Industrial Sciences (ENSAI), which trains the industrial engineers. In 2006, he founded Synergy Youth and Development (SYJEDE) that aims at achieving the Millennium Development Goals, particularly, access to clean drinking water. In 2011, he created a Green Technologies SARL, an engineering consultancy firm. He is currently working at the GIZ Cameroon as junior expert to support and implement the COMIFAC project for environmental and forest protection.

Philippe Quevauviller (France)

Philippe has been a researcher in oceanography and environmental chemistry before joining the European Commission, firstly at the Directorate-General for Research in the area of measurements and testing, then successively at the Directorate-General for Environment where he was in charge of EU water policy, back to Research and Innovation in the area of climate change and natural hazards, and now, Enterprise and Industry where he is in charge of research programming in the area of security. He is author or co-author of about 240 articles published in the international peer-reviewed literature, 50 book chapters and 60 reports; he is editor or co-editor of 27 special issues of scientific journals and 15 books, editor of two book series, and author or co-author of 6 books. He is also Associate Professor at the Vrije Universiteit Brussels (VUB) in the area of integrated water resource management.

Manjunath Sadashiva (India)

Over the last 23 years, Dr. Manjunath Sadashiva has worked in various volunteering and professional roles as a street children educator, citizen activist, visiting lecturer, trainer, manager, consultant and researcher. Two of his main professional interest areas are civic, political and environmental education, and democracy, civil society and participatory governance. He cofounded CMCA programme in the year 2000 and has been its full time Director since 2011. Prior to this, he was lead consultant on participatory governance to CIVICUS: World Alliance for Citizen Participation, Johannesberg, South Africa.

Lis Stedman (United Kingdom)

Lis is an associate editor of Water 21 magazine and was editor of Water Bulletin magazine. Over the years she has written for Water and Wastewater International, WET News, Edie.net and various other water sector and more general environmental titles. She is particularly recognized for her skills in writing technical challenges translated into universally accessible wording.

Acknowledgements

I would like to express my deepest recognition to the contributors of this book. I highly appreciate the time they dedicated to sharing their views or summarizing their scientific work and experience with water. They expressed their honour of taking part to this adventure when it should be reversed: what an honour to have them writing for this initiative. I thank you all for your dedication, your amazing work and your encouraging words about this project. I would like to thank my patient readers: Brita Forssberg, Maja Hodzic, Louise Hoogenhoot and Kyle Schneider. I would like to express my deepest recognition and sincere "*merci*" to Alex Bielak for his advice, understanding and time he dedicated to this project, it's been most helpful and encouraging.

I also take this opportunity to thank all the water professionals and citizens I questioned endlessly about their work and who are always keen on helping my research. Most of you are listed in the methodology (see Annex) but to those who didn't contribute directly to this work, rather having already participated to my other research projects, I'd like to thank you again for your dedication and answers about communication and water.

Preface

By Dr. Céline Hervé-Bazin

Water is life. Water is a right. Water is the oil of the 21st century. Universally, water is considered to be the most vital resource shared by humans, etc. These statements are consensual; they represent the most repeated discourses about water. Despite such common shared beliefs and public awareness about the importance of water, the world faces many kinds of water crises. Additionally, many water professionals consider the public has insufficient understanding of key, current water challenges, the reality of the water cycle and its technical requirements, its complex legal system or pricing. The long chain of impacts from our water consumption is clearly not fully understood. Many false beliefs circulate about water, leading to misunderstanding, fear, rejection, immobility, and misgovernance, etc..

Today, many hope communication can provide a solution. Information and Communication Sciences are a relatively new domain, particularly Environmental Communication, which has risen to prominence in the last decade or two.

This book provides an initial overview of the scope of Water Communication, by examining who communicates, and on what topics. I decided to organize our work based on geographical divisions. The approach is divided into the International, Global, National/Regional, and Local/community level. This choice is based on the fundamental characteristic of water as a local resource. Water is considered both a human value and a territorial resource. This scaled approach affords the opportunity to analyse the influence of different spheres, actors – organizations and target groups – and communities. As such, this book doesn't address the specific needs of communication by topic e.g. drinking water quality; water quantity; surface, ground, reclaimed, reused, recycled, purified waters; water governance; water rights e.g. legislation, directives, standards; etc. Many guides and scientific works have already been developed in these areas and we have provided some pertinent references related to those topics.

I hope this book will help you in your work, particularly to consider the impacts of your work and how to communicate it as you interact with different audiences. I provide two key elements to water sector professionals

1. **A select panorama of water campaigns** mentioning target groups, tools and scale. (See Annex 4).
2. **A reflection about developing communication campaigns on water** to assist your strategic choices in a globalized environment. Today, there are many contradictory messages and campaigns on water. By considering different messages, communication tendencies and resemblances, we want to help water professionals to understand how contradictory messages and campaigns can bring confusion and misunderstanding on water challenges today.

On a more personal level, I am attempting to set up a "communication water research group": Communicators in the water sector are a distinct minority when contrasted with what many consider the traditional composition of the water sector. This book represents a step towards greater visibility and integration of communication approaches with social and economic research in the water sector. Ultimately, I want to encourage more research, funding and integration of communication and social sciences in water management projects globally.

This book represents a first step in broaching such discussions. It requires further contributions, research and enhanced analysis of what has been done before. This work is based on my experience and therefore, referring to many examples from France. I am aware of the limits of this work but it will hopefully serve to identify research needs for more research on Water Communication. (See Annex 1 on methodology and reference material)

Please, feel free to contact us to share your views on this initiative.

Editorial concerns

This book provides several reproductions of logos and advertisements for the purpose of research and to support our description. Our objective was to give examples of existing campaigns about water management, not to use them for advertisements purposes or in a negative context and impacts. We did our best to gain the authorisations but despite our efforts some materials might be reproduced without appropriate consent. Please don't hesitate to contact us so that we can provide adequate corrections and amendments in future editions. Contact: celinehervebazin@gmail.com

Foreword to "Water Communication"

By Alex T. Bielak PhD

Founder and Chief Knowledge Broker, Alex Bielak Communications

With a number of interesting book projects and other initiatives already under her belt, Dr. Céline Hervé-Bazin has drawn on her burgeoning network of contacts in the "water world" to help supplement this treatise on water communications. Though I frankly doubt anyone would have been able to resist an invitation from this dynamic and persuasive communicator, it was nonetheless an honour to have been asked by her to provide the Foreword to such a broad-ranging work.

Clearly IWA Publishing recognizes the significance of this topic: to understand why I consider this book important, readers need some background. I hope they will indulge the diversion from a straight-forward foreword, and find my narrative both informative and helpful in understanding my enthusiasm for Céline's efforts.

Water permeated my life in many ways. Before I moved to *Waterdown* in Ontario, I lived in *Fall River* Nova Scotia. By the time this book is published we will likely be in the process of moving to the City of *Waterloo*! During undergrad studies in the seventies in Britain, my first work-term experience was with the East Suffolk and Norfolk *Water* Authority, and my second with The *Water* Research Centre in Stevenage.

My PhD thesis was on the salmon of the Quebec North *Shore*. I've had jobs with the National *Water* Research Institute at the Canada Centre for Inland *Waters*, as well as with the United Nations University Institute for *Water*, Environment and Health. I remain closely involved with the Canadian *Rivers* Institute and the Canadian *Water* Network,

So, I'm passionate about water and communications, and have had the good fortune to have worked, sometimes at relatively senior levels, on water-related issues with: Academe, NGOs, Foundations, Provincial and Federal Governments, the United Nations, and now, having founded my own communications company, the private sector. (Looking at it another way, I suppose I had a fairly short attention span when it came to jobs!)

You get the picture; it seems like it was meant to be. I always knew I wanted a career dedicated to water-related resource issues, but it is only with the ability to look in the rear-view mirror that I've come to recognize the common filament in my professional life was always communication.

My first work term in the seventies involved working with a commercial eel fisherman, water bailiffs, Senior Water Authority staff, the public and an English Lord. For someone not yet twenty years old, it was quite the crash course in tailoring messages!

Writing this foreword I've come to realize that this early experience was but the harbinger of what was to come. As the years passed, I found myself negotiating one-on-one with a First Nations Chief on an island in a remote river in Quebec for the return of PhD thesis samples taken during a tense occupation of a sportsfishing camp; persuading the Vice-President of an airline company of the importance of my studies to his business and the absolute necessity of giving me free passage when they had space aboard their planes; developing a series of public service announcements about salmon conservation, aired on major networks and featuring sports superstars (the eighties); and bringing together various constituencies when effecting major changes to fisheries and river management regulations (the nineties).

Later yet, in the first decade of the 21st century and working with an excellent and motivated team, I was privileged to help bridge science-policy gaps leading to major policy file changes via a series of Canadian Council of Ministers of the Environment-sponsored water-issue workshops, and develop broadly-adopted, science-policy linking tools.

The culmination of my official career was the conception of the KStar (K*) initiative and – as a United Nations staffer, in bringing the world's leaders in Knowledge Translation, Brokering and Mobilization together for the first time in 2011. I co-authored an analysis of the language of water in UN resolutions, help shape the communications strategy of the UN University as a whole to include consideration of knowledge brokering for the first time, and was also privileged to serve as the senior adviser to the Chair of UN Water.

It was not all roses along the way. At one stage I was asked to take on the job of raising the profile of the National Water Research Institute, Canada's largest freshwater aquatic institution. Between 2001 and 2005 my team did that systematically and successfully, to the point that we "owned" the top-ranked search term for the term "water research" on Google worldwide. The Institute garnered an U.N. "Science & Public Awareness Award" presented to the "…institution that has performed the greatest service to public awareness of fresh water issues in Canada through leadership in shared scientific research".

Despite achieving such extraordinary branding and recognition, and for reasons that are unclear to this day, but might make for a great episode of the classic British Comedy "Yes Minister," the NWRI moniker and distinctive bilingual logo were abandoned. During a chaotic re-organization NWRI was subsumed into, I hate to say it, the anodyne and bureaucratic-sounding "Water Science and Technology Directorate."

Vestiges of the old name still linger on the Internet because the quality of the scientists and their science was esteemed worldwide, but a California-based NWRI happily moved in to occupy the void as promotion of the Canadian institution was actively discouraged. Any marketer will tell you that the value of the brand that was essentially given away was inestimable, and you have only to look at the importance given to identity and water campaigns in Céline's book to understand why so many of us are still perplexed, not to say outright mad, at what happened.

In time I came to recognize that corporate communications in my Department was, then, largely about providing advice and never rocking the boat, rather than actually producing anything original themselves, let alone effecting change. Our group began to define itself by what it was not i.e. it was not performing a "Big-C" (Corporate) communication function.

Eventually – with the evolution of thinking about knowledge translation and brokering – we became defined by what our group actually was: i.e., a "little-c" communications group. We targeted our communications to specific audiences including scientists and decision and policy makers. By mutual agreement with the Big-C folk, we abandoned our previously-successful dealings with the media and the public.

Whereas Big-C dealt with "clients" I banished that word from our lexicon. We worked in partnership with others and I believe that was the foundation of our success, and the trust we built with our major constituencies. And indeed partnership is a term that flows throughout Céline's book.

We were so successful in our work that our mandate was expanded to include wildlife, technology and atmospheric research, where we continued to prosper and gain international recognition. However, with the ascendancy of a conservative government that did not appear to value evidence-based policy or open communication about science, we saw the fruits of good initiatives wither on the vine.

One particularly galling example related to our work with Water Conservation Authorities (CAs) in Ontario (Sheikheldin et al., 2010): It showed senior managers at the CAs were unaware of the volumes of pertinent federal science that could have helped them do their jobs better, and wanted to know more. Our cost-effective suggestions to broker such information were deemed unimportant by managers in the federal system running scared of falling afoul of political masters, or more interested in the bottom line than ensuring science flowed to those who could actually put it to use. So much for the science-policy interface, and a huge opportunity to have an impact missed.

Staff in the unit ultimately departed, were laid off, or eventually assigned new duties. In a final, ironic twist, in June 2013, my colleagues and I (about half of us were no longer with the Department by then) were pleased to be awarded a prestigious Environment Canada "Citation of Excellence" award in recognition of "outstanding innovation demonstrated in connecting people to science". The award recognized the creation and implementation of "EC Science Alert", a science-policy interface tool and was awarded in "the service to clients" category!

So in the latter part of a career threaded with water resource research and communication about water-related issues, I ultimately had the opportunity to reflect and develop my thinking about communications. In a 2008 publication (Bielak et al., 2008), written with colleagues across the world, we said it would be beneficial "to see a broader community of practice established to help bring people together." We proposed "the creation of a regular forum dedicated to knowledge brokering, where the community can meet and exchange information and experiences."

That thought lingered in my mind and led to my developing the global "KStar" (K*) initiative (Shaxson and Bielak, 2012). We ultimately defined K* as "*the collective term for the set of functions and processes at the various interfaces between knowledge, practice, and policy. K* improves the ways in which knowledge is shared and applied; improving processes already in place to bring about more effective and sustainable change*," and developed a spectrum of K* activities that are all systemically related to each other.

Figure 1 (below) – apart from its relevance to the section on knowledge brokering in Chapter 4 of "Water Communication"– is congruent and dovetails with the thinking in Chapter 1 where Céline provides a "glossary of key concepts to define (water) strategies and campaigns". These are: Awareness, Information, Communication, Knowledge, Mediation and Education, each of which could be neatly super-imposed in the K* Spectrum schematic.

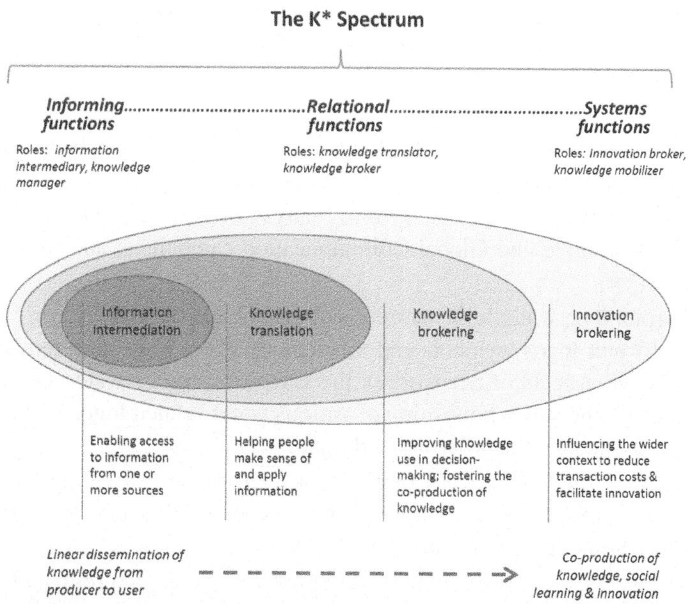

Figure 1 The K* spectrum. There is a spectrum of knowledge sharing activities, which are all systemically related to each other. (Adapted from Shaxson and Bielak et al., 2012. for a publication in submission by Shaxson, Bielak, Yip et al.,).

I freely admit to a bias for little-c communications approaches like knowledge translation, brokering, and mobilization, over Big-C Corporate communications approaches and campaigns, often involving the media. While I recognize the value of both, and that roles and functions can blur depending on organizational context, it seems the lion's share of funding has always been allocated to the latter. I've often argued that the balance needs to shift resources to include more targeted communications involving partners more directly. In my experience – and with notable exceptions like the brilliant work of the Water Supply and Sanitation Collaborative Council (WSSCC) which I've admired and encouraged Céline to include in the book – K* approaches can facilitate innovation and help it flourish, while a Big-C culture may stymie it.

The same applies in the world of water communication in my view. Billions of dollars have probably been spent on the sorts of flashy public and media campaigns showcased by Céline, but only a fraction of that figure might have been allotted to help broker information among water constituencies like science and policy, industry and communities, north and south, etc.

And that's where the bang for the buck (or Euro) often lies, I suspect, though both little and Big-C practitioners have not done nearly enough to properly consider dissemination strategies before they begin their projects or assess impacts of their efforts once the campaign is over or report published.

It is heartening therefore to see growing emphasis, in Canada, Europe and elsewhere on knowledge mobilization and improving research dissemination and adoption/application. The work of the Canadian Water Network and the European Water Community (e.g. the WaterDiss project) are exemplars in this regard and readers of "Water Communication" would be well served to reflect on such considerations as they peruse it.

Ultimately, you get the picture. I believe I've seen and experienced a fair bit in the water communications-related world. I've had some successes and suffered some knocks. And then I met Céline.

With the fearlessness of youth, she had a big idea, and didn't let its ambitious scope stand in her way as she ploughed ahead. She interviewed dozens of people, conducted surveys via the net, and pulled materials from a variety of sources to write a book, in her second language, no less.

As she says in the preface she ultimately wants "to encourage more research, funding and integration of communication and social sciences in water management projects globally" and the book - often, but unsurprisingly, illustrated with French examples - "represents a first step in broaching such discussions... (and) needs further contributions, research and enhanced analysis of what has been done before."

What I have found stimulating as I've interacted with Céline, providing input on earlier drafts, is the breadth of content and context she's considered: from Plato and Aristotle to Matt Damon and Bono; from individual initiatives to Corporate and Global Communication campaigns; and from water linguistics to the visual branding of water.

The sections in Chapter 3 on water lexicography and the colours of water are original and fascinating, as is her assemblage of water logos, symbols and information about high-powered branding campaigns. One of the truly innovative aspects of her work is the categorization of water as: natural, accessed, an element and drinking water: the super-imposition of various water-related logos on that schematic affords us a new perspective of the scope of activity in the domain.

Clearly, given what I've told you above, I'm glad to see knowledge brokering feature as an important category in Céline's spectrum of water communication, along with: Social Representations of Water: Water Discourses; Water Strategies and Campaigns; and Water Journalism.

I'm also thrilled she wants to convene a water communication research group and my experience with pulling together K* should give her encouragement, despite what might seem like long odds at the outset. Many would agree there is a plethora of overlapping, often useless, and sometimes competing water-related events, campaigns and initiatives world-wide. Some evidence-based, and, dare one say, complementary thinking on future water events and communications campaigns would be a huge step forward: as would assessment of the effectiveness of such campaigns and information dissemination strategies…

I hope readers are as stimulated as I was to think about different facets of water communication as they read this book. The thoughtful contributions by authors from ten or so nations to this book, and the inclusion of so many references and links, also contribute significantly to making this an important and enjoyable addition to the world's water literature.

I can already see a second, expanded edition as a twinkle in Céline's eye.

Alex Bielak
Waterdown, October, 2013

Main Results

Water Communication is a generic association of terms to designate all communication processes related to water as a natural resource (physical good) and a human resource (including services, uses, perceptions and beliefs).

Water Communication is built on a variety of academic disciplines and is at the centre of many scientific contributions. Water Communication can only be defined from various concepts, theories and scientific fields. Water Communication is the result of communication processes and interactions between different actors, discourses and social representations. Many guides provide concrete guidance's to help professionals to set up communication campaigns applied to the water cycle management (see Annex 3).

Water Communication is characterized by:

- A visual identity with the predominance of blue for the colour, the drop for the shape and the wave and circle for the dynamics.
- A limited use of words and linguistics resulting in a focus on water as a chemical and natural resource rather than a resource at the core of the actual civilized society.
- A homogenized rhetoric on water challenges sharing similar codes with Environmental Communication; abundant (often contradictory) references to ethics, responsibilities and public goods with an emphasis placed on uses.
- The three main opposing water discourses which have brought water cycle management to its current state of complexity: the relation to drinking water quality; the relation to water infrastructures; and the relation to Right to Water. These debates highlight oppositions between uses and users.
- The opposition between the messages, senders and historical discourses which have shaped the repetition of standardized campaigns characterised by the repetition of universal slogans: 'water and life', 'water for all', 'water kills', 'water is limited', 'water wars', 'water is a human right'.
- The fragmented nature of the water sector's identified senders: Each group of senders use specific type of discourses, messages and slogans. This culminates in a sphere of contradictory messages where individuals have difficulty in understanding messages, expertise and water ethics.

They are five main research areas for Water Communication: (1) Social Representations of Water; (2) Water Discourses; (3) Knowledge Brokerage; (4) Water Campaigns; and (5) Water Journalism.

The diversity and volume of discourses highlights the need for further research on communication about water cycle management and social approach in the water sector.

Successful communication on water cycle management should be based on the understanding of the scale, the target groups, wording and visual identity, as well as a thorough long-term vision and strategy employing different types of campaigns. The choice of messages, slogans and discourses is of high importance and requires the analysis and understanding of public perceptions, impacts on public opinion and the further integration of social approaches to water cycle management.

1 Scope of Water Communication

What is "Water Communication"?
Does it really mean something?

The word "water" commonly refers to fresh water and rarely suggests seawater resources. "Water Communication" applies to all forms of communication on fresh water resources. It includes processes, interactions, discourses, messages, logos, campaigns, codes and rituals disseminating information, symbols, perceptions and values on water resources.

Today, communication on water is recognized as a key element to planning, implementation, and operational decisions to optimise water resources and water cycle management. According to the IWA Specialist Group on Public and Customer Communication, "Communication is the key to successful water management". This recognition is the result of multiple processes influencing the water cycle management e.g. the incorporation of economics, politics, governance, and social sciences. As a realm of social sciences, information and communication sciences can be legitimately combined with water resources management.

Information and communication technologies have transformed society and the world in many ways. The appearance of mass media accelerated the development of new methods, new interfaces and new interactions. Information and Communication Sciences cover many different subjects. They combine aspects of both social sciences and the humanities, with influence in many domains. It is interesting to note that until today, information and communication sciences in the water sector were poorly understood and integrated as a key component of successful water management policies. This is a result of the relative youth of information and communication sciences but also due to the long tradition of water management based on technical and engineering competences; the insufficient integration of social aspects and the scarce amount of social sciences research relating to water management. Today, the terms "information" and "communication" are used as tools or materials rather than processes, strategies and interactions.

Information and Communication Sciences in the water sector have their own specifications based on codes, organisations, discourses, slogans, logos, and campaigns. The tracing of such conventions will help to further implement adequate decisions and programmes to optimize the water cycle to satisfy human, environmental and global development and welfare.

1.1 INFORMATION AND COMMUNICATION SCIENCES, A SHORT DEFINITION AND GLOSSARY

What are communication sciences?
What are information and communication sciences? Are they different?
What is communication?

"Communication" is a word with a rich history. Its definition dates back to the first interactions and has evolved with communication technologies, particularly mass-media. Concurrent with the spread of contemporary forms of transportation, modern media spread rapidly over the 1800s and 1900s. As described by Starr (2004), the development of media has been facilitated by several factors such as the state control over the public, the race for new technologies, economic development and political supremacy. The expansion of the industrialised world eased the process of implementing mass-media into the habits and daily life of the public. From newspapers and the print press to the more recent extremely rapid commercialization of web-connected phones and Internet connections, mass-media has dramatically entered the usual practices of our modern society in less than a century and a half (Herman and Chomsky, 1988; Sloan, 1991; Flichy, 1991; Jeanneney, 1996; Georgakakis and Utard, 2001). This does not include the latest major changes brought by social media, games and new web developments, impacting traditional media and society (Rebillard, 2009; Bouquillon and Matthews, 2010). Information and communication sciences originated from the development of information and communication technologies. However, they cover a larger spectrum of domains and concepts.

According to Mattelard (1994 and 1995), information and communication theories originated from the development of a philosophical approach to social organisation and the importance of discourses during the late 19th century, in particular the promotion of a global "utopia"[1]. According to Mattelard, the recent civilized society is built on the idea of a universal bond between humans, a shared world citizenship promoting peace and a homogenised vision of society.

For Mattelard, the enlightenment thinkers promoted those social, political and moral aspects based on the idea of Human Rights for all. Those shared beliefs initiated a global discourse. This global discourse is a process of communication. It unified people around one idea and supported the development of our modernized society. The influence of the Philosophers from the Age of Reason organized the birth of the industrialized world, the industrial revolutions and the spread of information and communication technologies (Mattelard, 1994). Those ideas expanded thanks to the development of printing technologies and telecommunications. Print media progressed with the growth of newspapers, books, pamphlets, comics, billboards and placards during the 19th century. Telecommunications appeared at the end of the 18th century and rapidly facilitated the transmission of messages, conversations and information broadcasts with the appearance of the radio in 1896. The development of films starting in 1880 contributed to the dissemination of this "Global utopia". According to Mattelard, it triggered and channelled socially shared discourses and references. This process is at the origins of information and communication theories and sciences. Information and communication sciences are organized around and study the processes promoting universal bonds, interactions, rituals and discourses towards individuals. They are key components to the establishment and the stability of industrialised societies. Mattelard reminds that communication sciences are an interdisciplinary field.

Interdisciplinarity involves two or more academic disciplines. Information and communication sciences usually interlap with philosophy, sociology, anthropology, psychology, biology, political science, economics, engineering and architecture. It attempts to study a given phenomenon in terms of multiple traditional disciplines. Information and Communication Sciences approach social interactions using several methods aiming to connect the different fields. They particularly consider the exchange of messages, construction of discourses, social interactions and relationships with information and communication technologies.

In 1927, Harold Dwight Lasswell was the first to conceptualize a communication model defining the origins of mass communication (Lasswell, 1927). Lasswell underlined the technical supremacy over individuals implying manipulation through the dissemination of messages. He further developed his model of communication in *The Structure and Function of Communication in Society* (Lasswell, 1948). The Lasswell model is the first to organize the study of "the communication process". His work is usually summarized by the famous sentence: "Who (says) What (to) Whom (in) What Channel (with) What Effect". His different studies of propaganda and communication during World War II produced new views on the impact of politics and their abilities to shape public opinion.

[1]A utopia is ideally perfect place, especially in its social, political, and moral aspects.

This approach of manipulation and propaganda is further widespread through a vision of communication tools as being functionalist, only sending messages in various types of channels. The most famous functionalist model is the Shannon and Weaver Mathematical Model (Shannon, 1948; Shannon and Weaver, 1963). This model is considered to be the founding model understanding the transmission of messages. This model is also called "the information model" or "transmission model". The model shows that messages are sent to individuals but does not consider feedback or impacts on individuals. Shannon and Weaver underlined technical constraints to the transmission of messages but conceptualized the transmission of messages as a unilateral and mechanical process.

The idea of influence, manipulation or propaganda is further developed in information and communication theories. Culture industry theories (Adorno, Benjamin, S. Hall) are interesting to mention. They showed the influence of cultural industries, including cultural programs and products disseminated on individuals and groups' identities. This mass-media theory criticizes individuals' behaviours and enhances domination by mass production, setting the roots of mass culture and mass consumption.

Human perspectives and actions are introduced during World War II, notably when Paul Lazarsfeld published *The People's choice* (Lazarsfeld, 1944). The two-step flow theory is the first contradiction of overpowered mass media. Lazarsfeld recognizes the influence of media, in particular the radio, on public's opinion and electoral behaviours. However, he introduced the importance of human relationships. Lazarsfeld highlights the role of opinion leaders or peers on individuals and within a community. According to him, opinion leaders are influenced by media; and opinion leaders influence individuals. This conception of the dissemination flow introduced the concept of social influence for the first time. Lazarsfeld showed the importance of processes, dynamics and influence. His work is the first step toward recognizing the individual's ability to interpret a message.

During the early 1950s, the School of Palo Alto initiated a new paradigm on the role and the capacities of individuals towards mass-media. They introduced a new approach to Information and Communication theories. They showed the capacity of individuals to respond to messages sent by mass media. They also pointed out the importance of "invisible" communication. The invisible components of communication processes include non-verbal communication, psychological factors, interactions, and interpersonal behaviours. Bateson, Hall, Goffman and Watzlawick are famous contributors to the work of the School of Palo Alto. They are also known as "the Invisible College". Their work is of fundamental importance in psychiatry and psychotherapy. They also contribute to define and establish social constructivism. In Information and Communication theories, the invisible college provided concepts and basics to study human behaviours and perceptions of mass-media processes. They started the concept that individuals or groups of individuals can be interacting, interpreting and using messages in a dynamic way. The Invisible College defined the concept of "feedback" that implies the return of information but also, the interpretation and return of information. It suggests a "second-order cybernetics". This process is more complex and calls for deeper analysis with personal, interpersonal and social appreciations. Individuals are the results of those interactions; they are able to build understanding of discourses, messages and campaigns. They are able to send back messages to the senders that will impact mass-media or cultural industries.

Information and communication processes are building interactions. Habermas (1962, trans. 1988) brought a radical new conceptualization of those interactions with the definition of the "public sphere". The public sphere is defined as an area where individuals come together to discuss societal problems, as a conceptualization of human interactions. Habermas conceived the public space as "a virtual space where the citizens of a country exchange ideas and discuss issues, in order to reach agreement about 'matters of general interest.'" Habermas described those interactions taking the example of the bourgeois society of the 18th century. He shows the impacts of rituals, hierarchy, distribution of wealth and the economy on the public sphere. The growth of the commercial society favoured the visibility of economic and private interests. The emergence of media accelerated the possibilities to manipulate discourses and interactions on the public sphere. This visibility impacted individuals who were losing their capacity to judge or to protect the general interest. During the 2000s, Habermas reconsidered individuals' participation and their critical analysis in his publication "the public sphere, 30 years after". The key contribution by Habermas relies on his conceptualization of social interactions and the role of individuals' judgement to protect human values and ideals.

In the late 70s and 80s, theorists further emphasized the roles of individuals and their perceptions. A series of authors brought new concepts to analyse both the construction of messages and discourses, and the reception by individuals.

- Austin, Wittgenstein, and Bakhtine contributed to establish the importance of linguistics to analyse discursive dynamics in communication processes.
- E. Hall and Mead further developed on the impacts of social and cultural interactions in the construction of communication processes. They established links between communication and anthropological studies.
- Foucault introduced the importance of power relationships in particular, from the perspective of organisations. He showed the role played by major institutions on individuals' behaviours and the organization of the society. His work on prisons has set many key principles on communication dynamics based on symbols, representations of power and discursive power.
- De Certeau analysed the capacities of people to transform mass products or communication technologies. He demonstrated the creative abilities of individuals to reinterpret or reuse objects, discourses or interactions brought by mass media.

Finally, recent communication research has focused on analysing the limits of communication discourses and potentials. Muchielli (2006) states that the word "communication" conveys promises of solutions whereas it should refer to processes and circulation between actors and organisations. Communication processes have their messages, their significations, their uses and impacts. This overwhelming influence of information and communication has contributed to illusions about communication processes (Jameson, Lyotard, Libaert). Others are proposing tools to understand the relationships between communication and individuals to consider those impacts (Eco, Wolton, D'Almeida).

Similar concepts to understand people's attitudes, behaviours, perceptions and actions are found in marketing studies. Information and communication Sciences overlap with Marketing when considering impacts and human behaviours. Several models offer an interpretation of behavioural changes. Today, organisations develop guides to explain such attitude transformations. They provide models that help to understand such processes. For instance, one of the main triggers for changing attitudes is fear or doubt. Ultimately the role of communication and program managers is to install a trust relationship (UNICEF, 1999).

Others will consider impacts of media and communication campaigns on individuals with regard to how media impact individual attitudes. Access to information and evidence will usually play an important role in changing opinions. Throughout the building of discourses (Breton, 1996), many researchers have investigated the relationships between the rhetoric of "proof", "mathematical systems" or legal systems in the current building of discourses (Mandressi, 2009; Herve-Bazin, 2012). They highlighted the importance of interfacing science with common knowledge.

This short history of Information and Communication theories and concepts underlined key principles and notions such as channels (technologies), senders/receivers, feedback, interactions and social interfaces. Today, Information and Communication Sciences want to follow, analyse and cover a growing spectrum of activities. Strictly defined, Information and Communication Sciences refer to the exchange of messages and their significations. More specifically, Information and Communication Sciences are a sub-discipline of the social sciences including human interactions, discourses or public interfaces. The discipline covers a wide range of application and research areas. These are necessarily interdisciplinary and use highly diverse methods.

Information and Communication Sciences became an institutionalized social science within the last 50 years. In the US, the National Society for the Study of Communication (NSSC) was created in 1950. In France, the French Society for Information and Communication Sciences was created in the mid 1970s. As such, the institutionalization and organization of Communication Science facilitated the development of new research on various themes and tools. It also accelerated the conceptualization of the different scientific perspectives and developed a history of Information and Communication theories (Boure, 2006 and 2008).

Information and communication sciences cover processes of producing, storing, transmitting, and accessing data and documents; relations between language and communication; interpersonal communication; intercultural communication, cultural practices; rhetoric and discourse analysis; mediation by different publics; mass media and journalism; impacts and uses by publics; organizational communication; political processes, discourse and their impacts. Recent developments of technologies fostered a growing number of research on computer-mediated communication, the specific content and impacts of new technologies in particular, the Internet.

Key fields include Information and communication theories; Cross-cultural communication; Development communication; Discourse and rhetoric analysis related to linguistics, socio- linguistics; Environmental Communication; Health communication; International communication; Mass communication; Media studies and Journalism; Organizational communication, including internal communication, external communication, crisis communication, management communication, public relations, institutional communication; Political communication including electoral communication, lobby and advocacy, influence communication and strategies, public opinion; Science communication including scientific interfacing, knowledge brokerage, risk communication, technical communication; Social communication and marketing and visual communication, including semiotics.

From this overview of Information and Communication Sciences, here is a glossary of key concepts to define strategies and campaigns (Box 1.1).

Box 1.1 Glossary of key terms in Information and Communication Sciences

AWARENESS

Awareness is the ability to perceive or to be conscious of events, objects, references, etc. It is the first step to acknowledge the existence of a given phenomenon, fact or group. Awareness campaigns aim to bring visibility to a phenomenon so it can enter common knowledge. Awareness campaigns should be considered as the very first step to approach a public on a specific topic. In the water sector, many awareness campaigns exist to bring visibility to water challenges. They are based on simple and powerful messages with images as opposed to information campaigns (data-based) and communication campaigns (values- oriented). Many are focusing on water uses (water conservation, water waste) or global changes (links with climate change).

Awareness campaigns are characterized by images, simple words and common references.

Box 1.1 Continued

INFORMATION

Information Science refers to an interdisciplinary field primarily concerned with the analysis, collection, classification, manipulation, storage, retrieval, movement, and dissemination of information. Information involves both knowledge management (knowledge centres, libraries), and the dissemination of this information, e.g. information campaigns.

Information campaigns aim to provide informative content. They do not intend to influence or provide analysis. In the water sector, information campaigns are important in providing statistics or legal references. Key topics are water quality (bathing, drinking standards, etc.), water quantity (water levels / water resources, water uses, meteorology), water governance (legal background, public bodies, projects, etc.), water economics (pricing), and water health (human body, diseases).

Information campaigns are characterized by data and information references.

COMMUNICATION

Communication Science addresses human and social processes, producing shared symbols, meaning and discourse.

Communication campaigns aim to produce cohesion, shared values and social references. They intend to create economic values, identification and assurance, and potentially, influence people's behaviours, attitudes and perceptions. In the water sector, communication campaigns are bringing messages, values and social shared actions related to water challenges. Key topics are water access (water development, water services, humanitarian water); water drinking (bottled water, tap water); water right (water ideology e.g. private / public water, UN objectives); water uses (domestic uses, cities and infrastructures); water research and technologies (desalination, irrigation, etc.); and protecting water resources from degradation by human activities.

Communication campaigns are characterized by values disseminated in slogans, messages, and discourse references.

KNOWLEDGE

Knowledge comprises facts, information, descriptions, skills, experiences providing understanding of situations and subjects. Knowledge involves dynamic processes, meanings and representations.

There are many sub-fields to knowledge, including knowledge management, knowledge translation, and knowledge brokerage. Knowledge management is related to information management, but extended to experiences and practical knowledge. Knowledge translation is the activity of adapting the content to specific language and references to address a group or community. Knowledge brokering is the activity of producing links between two groups of people, the producers and users of knowledge and giving them knowledge sources and "translated knowledge".

Knowledge campaigns aim at facilitating linkages, brokering and translation activities between two groups. It aims to provide tools and contacts to understand

Box 1.1 Continued

others' experiences and challenges. These types of campaigns will imply a change in attitudes. In the water sector, some projects developed the concept of policy and science interfacing. They suggest promoting links between SMEs and Utilities, NGOs and Local associations links, etc.

Knowledge campaigns are characterized by links, rituals and shared references.

MEDIATION

Mediation is a legal term and is used to refer to dispute resolution. Mediation involves dialogues, discussions and techniques to make two or several parties agree on a certain action. People or organisations entering mediation processes are usually ready to find solutions and to make compromise.

Mediation campaigns aim to resolve challenges related to water management. They intend to establish common references for actions or dialogues. These campaigns imply a change in behaviours, practices and attitudes. In the water sector, mediation campaigns could be undertaken for basin levels or for cases of disputed waters access rights.

Mediation campaigns are characterized by sharing and experience references.

EDUCATION

Education is a broad word encompassing various forms of learning. Education and Communication are complementary processes. The combination of Education and Communication implies processes, interactions and strategies to impact individuals' behaviours. They attempt to affect attitudes, beliefs and acts. They have long-term approaches and objectives. They are usually implemented at a local level or with specific target groups. Education and Communication Sciences are traditionally interconnected when it comes to Environmental Communication (biodiversity, Sustainable Development, climate change).

Education campaigns attempt to provide knowledge, experiences and interactions to change people's perceptions and understanding of societal phenomena. They imply a change in beliefs. In the water sector, education campaigns relate to the water cycle (water classes, iconography, experiments) and targets kids and teenagers. Several education campaigns exist on water pricing (how pricing is established), and on water history (water and civilization).

Education campaigns are characterized by long-term experiences and responsibilities.

1.1.1 In Short

Information and Communication Sciences cover a variety of subjects, methodologies and techniques. They aim to analyse processes, interactions and dialogues. They embrace activities from organisations and individuals. They look at messages, discourses, slogans, campaigns, codes, rituals, events and uses of technologies. They examine symbolic meanings and interpretations. They consider impacts from information and communication technologies, mass media, mass culture and mass products. They examine strategies and effects on people's awareness, knowledge, attitudes, behaviours, perceptions and beliefs.

Information and Communication Sciences provide keys to define strategies to address public opinion, and to form organisations' interactions. They have different impacts depending on media, strategies, tools and interactions.

The different stages of information and communication campaigns create different types of references that influence individuals and communities (Figure 1.1). These stages eventually create changes, the institutionalisation of community and shared responsibilities. Water cycle management represents a humanitarian challenge. Water is a natural resource; a vital and universal commodity. It demands responsibility. The importance of information and communication to ensure water cycle management reinforces this particularity. The links and relations between water and communication lead us to review the history of interactions and discourses on water.

Awareness	• Give the information *about the challenge*
Information	• Provide the information with details *about the challenge*
Communication	• Facilitate identification *with the challenge*
Knowledge	• Ensure the understanding *of the challenge*
Mediation	• Create a dialogue *on the challenge*
Education	• Bring changes and self experience *on the challenge*
Responsability	• One's decision to act and change *for the challenge*

Figure 1.1 Information and communication campaigns.

1.2 THE INTERNATIONAL RECOGNITION OF WATER COMMUNICATION

The recognition of a specific Water Communication is recent. It originated from a process initiated at the international level to improve water cycle management.

1.2.1 The Role of The UN

The United Nations (UN) recognized the importance of raising awareness and providing tools to communicate since the launch of the very first decades. The first decade, on Development, was launched in 1960. The dissemination of information (statistics) was part of the objectives of the Decade (UN Resolution, 1710 (XVI), paragraph 4. action h). The UN resolution did not include the public or communication campaigns. Activities were limited to the exchange of information. However, it represented a first step towards interactions and information management on water-specific data and activities.

In the late 1960s, the International Development Research Centre in Canada was one of the first to develop models for water and sanitation access, involving the participation of communities. UNICEF and many Non Governmental Organizations (NGOs) also started working in local communities. They initiated the first social methods as opposed to technologically-oriented solutions. UNICEF is leading in many aspects: the dissemination and implementation of information; the encroachment of communication into the field; setting visible communication campaigns to the general public. The need for the organization to raise funds explains why UNICEF is pro-active and innovative in bringing visibility to humanitarian work. In this regard, UNICEF is one of the first organisations to publish a complete guide on communication for water supply and environmental sanitation programmes (see Annex 3).

In 1971, for the second Decade on Development, the UN recognized the importance of mobilizing the public on resolution 2800 (XXVI). The Economic and Social Council declared the first Decade for Water, the "International Drinking Water Supply and Sanitation Decade," in 1979. The declaration also provided guidance on communication strategies:

> This note gives an overview of the unique challenges that professionals in the water sector face. It also provides some general guidelines and lessons to be aware of when designing and implementing a communication campaign. It is not a step-by-step guide for developing a strategic communication program. Rather, it focuses on the specific tools needed for governments and practitioners to better understand their customers, how to build a case for support, and how to build a culture of communication within the utility management and among employees (etc.).
> These principles include (i) getting to know your customer, or potential customer; (ii) building awareness around the need for reform; and (iii) building a culture of communication within the utility and between the service provider and the population it serves (see section 3).

The creation of the UNESCO-IHP, the International Hydrological Programme (IHP) in 1975 corresponded to a new step towards more dissemination of knowledge. In 1975, UNESCO launched the programme to create a centre of data and knowledge. The IHP serves as a centre to study and collect information on hydrological cycles (Sadler, 1987). In this process, UNESCO also started to initiate a new approach to traditional water management. UNESCO organized several seminars treating economic and social aspects of water (Glasser, 1975).

The IRC, the International Water and Sanitation Research Centre, played an important role in promoting new approaches to water cycle management. The organisation was founded in 1968. The IRC is a "knowledge-based NGO". The organisation recognised the importance of humans in water cycle management. The organisation is one of the first to conduct research and programs activities integrating methods from social sciences. Looking at their extensive library and work, the IRC has developed many social programs. In particular, they developed methodologies to encourage participation of local communities. During the late 1970s, the IRC promoted several studies based on communication approaches. Their scientific work integrated social interactions or the analysis of images and perceptions in case studies.

For instance, Christine Van Wijk-Sijbesma (1979) identified the importance of peers to the shaping of the community's opinion. She recognised that they are key parties to implementing successful water and wastewater programmes (Van Wijk-Sijbesma, 1979). Van Wijk-Sijbesma also pointed out the importance of mass media on opinion leaders. She argued that communication campaigns could work to change people's uses of water resources, while highlighting the importance of long-term initiatives. In this work, Van Wijk-Sijbesma clearly identified a two-step flow of communication (Lazarsfeld, 1927). She also confirmed the importance of communication campaigns and education campaigns. The work of the IRC shows that the integration of communication sciences dated at least, from late 70s. At that time, such approaches still appeared to be isolated.

The UN organised the first decade on water between 1981 and 1990. This decade was named the International Decade for Water Supply and Sanitation Decade, the "IDWSSD" or the "International Drinking Water Decade". The aspects of communication, participation and mediation started to be promoted. The initial preparatory documents dated from 1979-1981 show that the UN developed a communication strategy to promote the Decade. The slogan became "Water and Sanitation for All". Newsletters, events and guides appeared as well as the development of a dedicated logo for the decade.

This decade multiplied the number of publications, brochures and first campaigns on water at local and national levels. Many organisations participated, including UNICEF, UNESCO, the World Bank, NGOs and various research institutes. The collective published their intial brochures, books and scientific studies. The objective was to legitimate the group's model of development and approaches to water cycle management. Recurring characteristics constituted a format. Those organisations created the roots, codes and references when communicating on programs for water and sanitation access. They established a uniform communication construction, including:

- General data: the argumentation is built on alarmist numbers, calling for urgent actions (number of deaths, water access, required investments, etc.)
- Case studies: Different local studies conducted in different countries (preferably in the five continents) are showing concrete impacts, return of experience and the universality of challenges.
- Experts: Figures providing knowledge and confirming the legitimacy of the actions.
- Local communities: men or women are bringing their testimonies. They are also in pictures to personalise and show benefits for individuals.
- Steps/Guidelines/Recommendations: The content wants to provide guidelines to other practitioners.
- Advocacy: the style is institutional and policy oriented. Publications aim at convincing policy makers to invest more. They are also justifications of the invested funds.

The first Water Decade instituted many first initiatives and opportunities for UN organisations and water professionals to communicate. It created a movement to bring visibility, information, and exchanges of practices and interactions on water cycle management. It established the first discursive references, slogans and content management among water organisations. However, the decade was considered by many as a failure. It did not reach the public's attention or receive media coverage. It did not bring major actions and engagement by UN Member States. The first Water Decade did not result in global action.

The water sector remains intrinsically related to the public interest for environmental causes. During the 1980s, the public interest for the environment or natural resources was low. Governments are also not involved and rarely engaged. The second Earth Summit in Nairobi is a concrete example of a global failure of Environmental Communication. Environmental discourses took a more important place in the media in the early 1990s. At the political level, the organisation of the third Earth Summit in Rio in 1992 was a key moment for governments to express their commitments to better preserve the environment. The Agenda 21 set routes and implemented planning at all geographical scales. Communication on water challenges benefited from this general movement towards more visibility, interactions, discourses and campaigns.

The acceleration of the number of conferences on water and, in particular, the creation of the World Water Council in 1996 participated to spread activities, communication and lobby on the water cause. The World Water Forum organized events in The Hague (2001) and Kyoto (2003) that gathered more than 20,000 participants. The two forums brought more communication about World Water Actions at the Global level. The concepts of a Blue revolution and World Water Vision participated to unite the sector into one coherent set of communication terms, platforms and global actions.

The combined work of the UN "Mega-Conferences" (Biswas, 2004) and the specific actions of world water organisations clearly contributed to the emergence of a water discourse that accompanied the development of codes, logos, slogans, visions and shared discourses specific to the water sector (Biswas, 2004). In this sense, framing and naming water priorities is part a multi-actor influence, participation and the construction of rhetoric and discourses (Mount and Bielak, 2011). As such, it constitutes the important keystones to the constitution of "Water Communication" itself. The number of events, the creation of rituals (events, prizes), the specificities of programs and organisations contributed to the ritualization and "sloganisation" of the water sector (Tournier, 1985).

In 2000, the UN launched the Millennium Development Goals (MDGs). This global action clearly stated the importance of better communication on environmental resources, including water resources. The 2003 International Year of Fresh Water raised visibility on water challenges. The UN clearly identified the challenge of communication during this year. As a result, UN-Water was created in 2003. This new UN organization is a response to the challenge to improve information and communication on water challenges. The UN-Water acts as a coordinating mechanism, inclusive of communication management. The UNW-DPAC program, the UN Water-Decade Programme on Advocacy and Communication aims to develop communication campaigns. This coalition highlights the institutional recognition for dedicated communication on water-related issues.

The year 2003 was also the International Year of Water, the year of the World Water Forum in Kyoto gathering more than 24,000 participants and the announcement of the second Water Decade (A/RES/58/217). In this general context of an increasing visibility of water issues, the production of communication materials, activities and the collection of information and best practices were multiplied. Starting in 2003/2004, many communication strategies for the water sector at national levels were widely developed in many countries, particularly in Africa. The creation of the UN-Water organisation supported the coordination of produced content.

On the website of UN-Water, the organisation is defined as "a coordination mechanism for all water-related issues" and "is responsible for coordinating the 'Water for Life' Decade". In several aspects, the declaration to launch the International Decade for Action "Water for Life" in 2003, includes social and human aspects:

> "Water for people, Water for life" as the title of the UN World Water Development Report printed in 2003
> Consider water, sanitation and human settlements as the thematic cluster of the first cycle 2004–2005
> The phrase "for life", which refers directly to humans, animals and nature.

Interestingly, there is no direct mention of individuals, or the public; stakeholders appear to be too recent a word and concept to be included in the official text. However, women are mentioned: "while striving to ensure the participation and involvement of women in water-related development efforts" (UN Resolution 217, 2003). This recognition, continues the Declaration of Dublin, specifies gender aspects rather than public participation and is part of building the discourse over water and women (see Chapter 6).

The second decade for water started in 2005, it introduced the importance of communication as part of its main activities. One of the UN-Water programmes is entirely dedicated to communication: the UN-Water Decade Programme on Advocacy and Communication (UNW-DPAC). The programme specifically addresses non-expert populations and media.

> "The mission of the UN-Water Office as the UN-Water communication hub includes coordinating communication actions of UNWater and its members and partners while at the same time implementing specific communication and advocacy activities towards achieving the Millennium Development Goals and those of the Johannesburg Plan of Implementation." (Biennial report, UNW-DPAC, 2011)

Through this decade, the office of the UN is organizing global events and awareness-raising campaigns until 2015. The aim is clearly to accompany better water and sanitation management.

> "The role of the Office in supporting UN-Water is fundamental, as UN-Water recognises that advocacy and communication are key for raising awareness amongst governments and stakeholders of the need to achieve international agreed goals on water and sanitation." (Biennial report, UNW-DPAC, 2011)

Posters and conferences on water and communication originated by UN-Water can be found starting in 2008. The creation of different specialist groups and training programs reveals that the subject is now considered to be a specific scientific area of expertise. During the 2000s, activities sought to address marketing and behavioural topics (IWA, USAID); increase water education (UNESCO-IHE, UNESCO-IHP); or encourage Journalism within the water sector (UNESCO, UN-Water, RNTC Multimedia Journalism and Water).

Other national organisations and UN organisations have produced materials and programs to increase links, activities and awareness on water and communication. For example, in Canada, starting in 2007, several programs targeted journalists or the public on specific topics, such as the Great Lakes.

Overall, these activities influenced the production of brochures, campaigns, and events starting in 1979. There is a clear rise in the number of specific brochures dedicated to guiding professionals toward better communication on water over the last 10 years. The number of organisations, publications and brochures in addition to campaigns at the global level impacted specific skills, practices and discourses on communicating water issues.

1.2.2 The IWRM, A Paradigm Shift to Water Management

This institutionalization of research on communication about water issues is related to the increase in research on environmental resources, aiming to improve communication to different target groups. It is also related to the institutionalization and dissemination of research on water management and Integrated Water Resource Management (IWRM). Research and publication on technologies and management from hydrology and engineering are the most predominant approaches (Biswas, 1970; Levi, 1995; Dooge, 2004. See references in Bibliography, III. History of Water)

The approach that IWRM initiated between the 1920s (Embid, 2003) and the 1930s (Snellen and Schrevel, 2004) introduces other dimensions to water resource management. The first attempts to implement this approach were defined by Pearse et al., in 1985 for the Canadian government (Pearse *et al.*, 1985). Those guidelines considered taking "all uses of water" into consideration. Pearse et al., also developed the approach of "watershed", an analytical system, and modules as well as specific management requirements, involvement of regulatory agencies and the participation of the population.

Today, IWRM's definition has been accepted as "a coordinated, goal-directed process for controlling the development and use of river, lake, ocean, wetland, and other water assets." This definition by the Global Water Partnership includes five key principles:

1. Water is a finite and vulnerable resource.
2. The implementation of a participatory approach.
3. The importance of the role of women.
4. Acceptance of the social and economic value of water.
5. Integration of the 3 E's: economic efficiency, social equity and ecosystem sustainability.

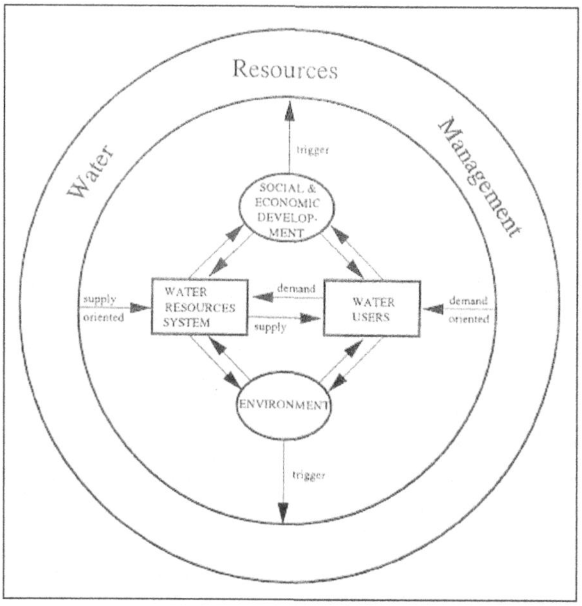

Figure 1.2 IWRM, a conceptual model. In this figure, Snellen and Schrevel (2004) show how that the IWRM is based on key communication principles: interaction, cycle, different groups exchanging information and a global context symbolized by the circle showing that any initiative for water management is part of a specific world/sphere.

As one can observe, there is no direct consideration of the communication approach in the fundamental elements of IWRM (Figure 1.2). IWRM introduced the participatory approach to traditional water management and sciences. Participation involves reaching the public, individuals, stakeholders, etc. Participation admits that technology is not self-sufficient to resolve challenges with water cycle management. A participatory approach includes interactions, the understanding of uses, and the analysis of perceptions. IWRM is a conceptual approach that opened water management to social sciences, in particular, to Information and Communication Sciences.

The other important pillar is the idea of "a coordinated (etc.) process". The concept is introducing key elements that were not acquired in the traditions of water management: public involvement (participatory approach) and interactions with other actors that later led to the notion of "stakeholders". IWRM states and values the role of users, of participation and of processes. IWRM plays a key role in promoting communication as an approach based on information and communication concepts. IWRM introduced a paradigm shift to water sciences that has allowed for the development of specific communication processes for water management in research.

1.2.3 The Emergence of Research on Water Communication

Looking at scientific research related to the spectrum of communication on water, a great amount of research could be found in various titles under different social science approaches: marketing, including behavioural change; perceptions, representations and anthropological origins; discourse analysis, media coverage.

The research appears to be fragmented. The observation of fragmentation is consistent with the criticism of the water sector as being fragmented and lacking coordination. This fragmentation can be considered as the result of the "pump-and-dump mentality" (Robert, 1993). In his work, Robert also points out that this lack of global vision can have substantial impacts on water management and water resources (Bohme, 1988) and is contradictory to the nature of water.

> Water flowing essence is to be elementary and ubiquitous, is institutionally channelled along tracks corresponding to disciplinary divisions in a society dominated by professionalism and its classificatory mentality. (Robert, 1994)

Social aspects have been otherwise studied through research on anthropological aspects of water starting in the late 1950s (Singer *et al.*, 2011), the analysis of public uses of water and relationships with water sanitary qualities (Goubert, 1986) as well as with social representations of water originated from water symbolisms and significations (see Bibliography IV). This legacy will be later developed, inspecting the influences from other communications fields.

When looking in a dictionary, the expression "Water Communication" exists, but only applies to transportation[2]. When searching on the Google in September 2013, results will show references to "under Water Communication" or "water conductivity". There are very few mentions of communication on water related issues (Blog in Canada mentions the need to make water more visible or results will link to guides mentioned in Annex 3). They usually will not describe communication activities performed by organizations from the water sector and/or performed by water professionals. As such, the expression primarily refers to the transportation or exchange of fluids. This understanding of "Water Communication" can actually drive back to the first theories of communication (Shannon and Weaver, 1963). Water Communication could be seen as a medium, a channel, and a device for transmitting information and influencing people rather than a communication process. This could be the first acceptation for a definition of Water Communication: water influences the minds of the individual and the way they interact with it[3].

The second step would be to define "Water Communication" as the result of processes and interactions. In this frame, the association of the two terms appeared more recently. This apparition is related to the growth of research and studies on Environmental Communication. Water is a natural resource. Water is included in the scope of Environmental Communication. As a "sub-topic" from Environmental Communication, Water Communication will have the same characteristics.

A third step would be to integrate the particularity of water as a natural, vital and social resource. If considering water as a transversal good, research on Water Communication could be found in communication research about nutrition and health from the agro-industrial perspective; political communication; public and territorial communication or communication on laws.

From this preliminary position, Water Communication is built on a variety of academic backgrounds and is at the centre of many scientific contributions. Water Communication can only be defined from various concepts, theories and scientific fields. Water Communication is the result of communication processes and interactions between different actors, discourses and social representations.

Many different organisations published guides and manuals to implement communication strategies and campaigns. As mentioned, UNICEF published a comprehensive guide in 1999. The book proposed different steps to set up the program's strategy using different methodologies from sociology and communication analysis. It also provided different case studies to show the variety of issues at the local level such as

---

[2]Including a special mention from 1830 "Canada, Water Communication", a book about transportation from the Parliament of Great Britain.

[3]Shannon and Weaver defined communication as "all the procedures by which one mind may affect another. This involves not only written and oral speech, but also music, the pictorial arts, the theatre, the ballet, and in fact all human behaviour" (Shannon & Weaver, 1949, p. 3).

health and water quality; sanitation, sustainable water management; micro-credit; etc. The authors refer to communication to establish their conceptual approach. Their main objective is to initiate new behaviours that are sustainable throughout time. Communication is considered a way to reach social transformation. Social mobilization, advocacy, community participation, programme participants and programme communication are key concepts to reach objectives. The guide suggests a model of Communication for Development as well as key steps to follow for successful implementation (Figure 1.3).

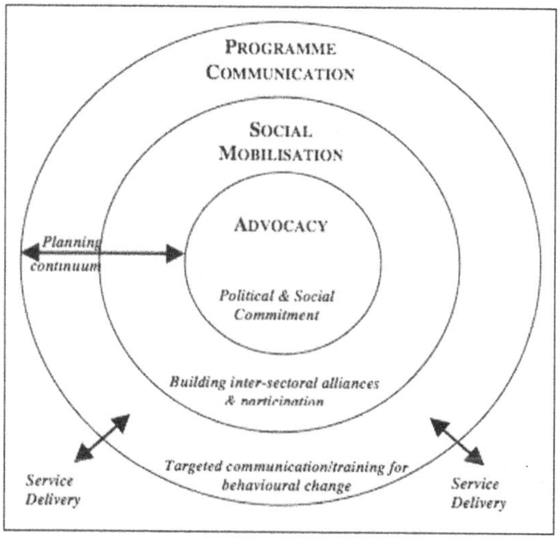

Figure 1.3 Communication for development model (Source: UNICEF, 1999). The model is based on the assessment of the situation (what do we know?), the identification of programme participants (who are the target groups) and how much they are likely to get involved (why would they join); the assessment of best channels to provide them messages (how to do we communicate); and the design of a strategic plan (when, where, with whom in order to set up the appropriate timing, roles' repartition and the number of activities.).

The US Environmental Protection Agency (EPA) has proposed practical and conceptual approach to communicate "the value of water". This book is aimed at utilities; it provides background studies and analysis to deliver successful messages about water to the public. It states that one of the first needs and rules for water utility managers is to "communicate, communicate, communicate! (If you can't, get someone who can to help you)" (EPA, 2008: 4). The main idea is to communicate "the value of water". The study recalls the definition of value as "the quality of something that makes it more or less desirable, useful, and a thing of quality having intrinsic worth." The appreciation of value depends on many factors including "cultural attributes, spiritual and emotional values, environmental issues, recreational issues, economic development and sustainability issues, institutional ownership issues, water supply reliability, infrastructure reliability, water quality, aesthetic issues, and other concerns (EPA, 2008: 4[4]).

According to the EPA, the main objectives of communicating the value of water are to convey the value of water itself and the value of the public services associated with the management of water resources. To communicate about water is related to the sense of public service and public good. The communication aims at resitting a sense of community. The EPA also postulates for the need to explain water management services, the cost of infrastructures, the need for efficient and effective management of water resources. It

[4]Quoting Raucher, R. S. (2005). & Raucher, et al., (2005).

provides steps to communicate the value of water in order to set successful communication and strategic plans. The EPA defined eight key steps (Figure 1.4):

1 Evaluate the existing situation and future scenarios
2 Establish the desired utility brand and communication vision
3 Develop communication goals
4 Identify needs and gaps that must be addressed in order to achieve the communication goals
5 Establish current utility behaviour
6 Identify behavioural needs and gaps that must be addressed to achieve the goals
7 Prepare a communication plan and integrated it into the utility's strategic plan
8 Implement and monitor the communication and strategic plans

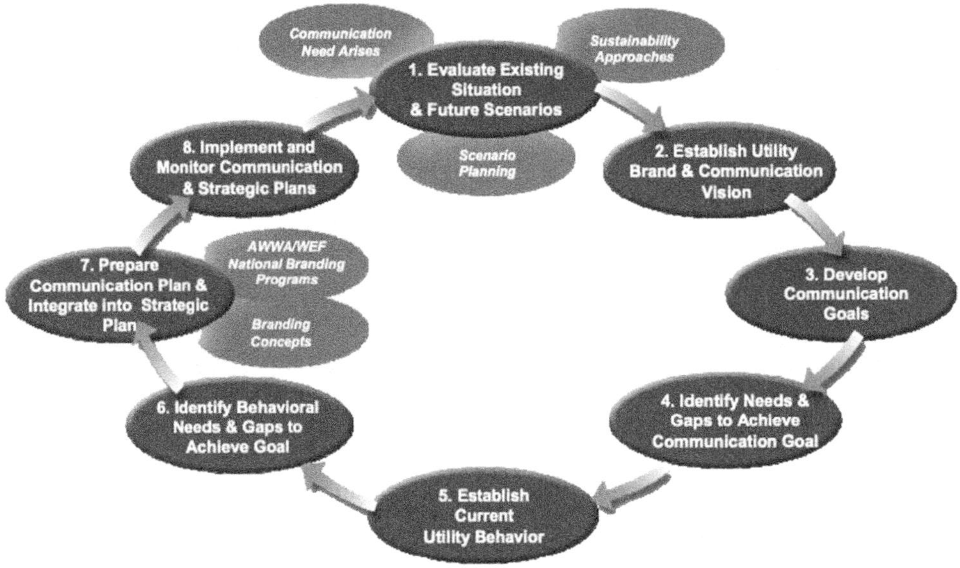

Figure 1.4 Model for developing plans to communicate the value of water (*Source*: Means (2008, p. 100)). The model showed the importance of strategic planning, assessing perceptions and behaviours, to strategic planning. The model particularly highlighted the importance of vision and the involvement of the top management.

The study by the EPA confirms the importance of understanding the stakeholders' needs and perceptions, the influence of values, the context of the communication plans and objectives, the choice of media but more particularly, confirms that Water Communication is a specific process. They identified several ideas, opportunities and research areas to better asset the needs and perceptions of stakeholders and to better "communicate the value of water". According to the EPA, the main areas to further develop are first related to the consumers: the consumers' perceptions of the water services in particular, perceptions of trusts – and risks; consumers' needs and uses to develop adequate water services; consumers' attitudes towards major challenges (environment, Sustainable Development, etc.); consumers' perceptions, reactions and their education about extreme events. The other main area is about information and existing materials. There is a clear need for databases, examples and guidance for utilities but also for media and the public. This study confirmed the gap between utilities, their constraints and their target groups. The idea of "communicating the value of water" implies the concept of particular codes; needs, perceptions and attitudes towards water that should be further researched and understood. It also suggested that the need for such a communication was not yet fully understood and integrated by utilities in their daily practices. The study acts both as a guide and advocacy arguments.

The UK Department for Environment, Food and Rural Affairs (DEFRA) produced a study for "Drinking Water Quality Communication" in 2009 (prepared by Risk Solutions, 2009). The study showed that the need for guidance's on communication about water was not strong. The main objective of the study was to share practices and ideas. The authors particularly underlined the importance of communication processes. With a similar methodology to the one of the EPA, they offered ten key stages when communicating with customers and agencies (Figure 1.5). In the model by DEFRA, the ten stages include:

1. Established communication channels
2. Prior allocation of roles
3. Advance preparation of materials
4. Internal alerting to planned event
5. Recognising and receiving the problem
6. Internal alerting to unplanned event
7. Classifying & escalating events
8. Selecting, preparing and sending the message
9. Debriefing
10. Training and testing

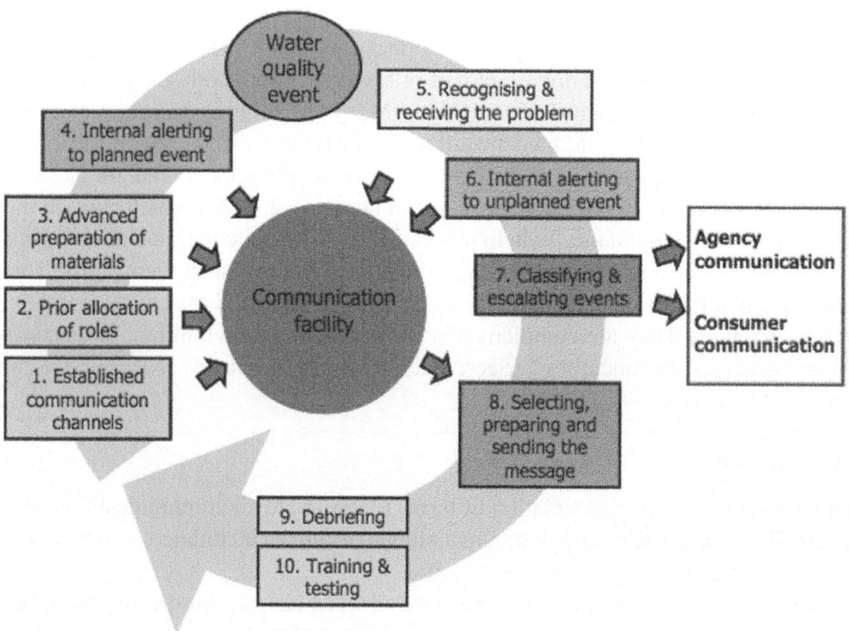

Figure 1.5 Water quality communication life-cycle (Risk solutions, 2009). The model is applied to issues related to water quality e.g. information and water pollution event. It is also designed for two key target audiences: customers and agencies. The model shows the importance of preparation (strategic planning), prioritization and assessment of communication processes.

The guide also shows the relevance and professionalism of communication campaigns by water companies. There is no doubt; water companies and stakeholders offer a wide range of successful communication campaigns (see Annex 4). There are many innovative communication campaigns about water every year and many have received prices. However, the importance of sharing practices remains an issue as well as a better integration of communication into water services' activities. The British guide acts as its American counterpart: both are guides providing advocacy arguments.

In 2012, the World Bank issued a similar study and guide for communication in the water sector. It argues for the building of "a culture of communication within the utility management and among employees" (World Bank, 2012: V). The study also provided steps to communication on water in the context of "water sector reform". The model is also applicable for utilities, policy makers and service providers. The model includes three key steps:

1. Get to know existing and potential customers
2. Build awareness around the need for reform
3. Build in support and systems for change

The World Bank developed this work to accompany programs aimed at developing water access. The study proposes general information about strategic communication along with case studies on water in different countries. The World Bank argues for a communication process that will ultimately "help revenues and make utilities sustainable". The guide is also acting as a guidance and advocacy argument. The challenge of affirming and integrating communication on water as a specific process remains an issue. There is a need to better integrate communication and to integrate communication processes that are relevant to water cycle management and public's perceptions.

Several research centres produced the study of global campaigns. The campaigns developed by the UN and for major events represent an important part of research. On the other hand, many brochures and researchers have analysed local and national campaigns. Many have shown the importance of multi-media campaigns or in certain area, they highlighted the influence and impacts depending on the media, topics, and interactions. Many for instance, showed the importance of radio at the local level, in particular, in Africa or South America. Others have shown the importance of combining media to better reach their audience. Many guides provide concrete guidance to help professionals to set up communication campaigns. Those publications provide tips and tools to facilitate the implementation of such campaigns. They offer very practical approaches to information and communication campaigns with different case studies or examples of successful best practices. They are complementary in considering the building of discourse on "water campaigns" and "Water Communication". (See Annex 3) Many examples of successful campaigns are provided throughout this book.

1.2.4 In Summary

Water Communication is a generic association of terms to designate all communication processes related to water as a natural resource (physical good) and a human resource (including services, uses, perceptions and beliefs).

Water Communication suggests that communicating on water has its own specificities related to the fact that water is a unique resource. It considers that water is different from other environmental, sustainable or renewable resources such as wind, air, soil, etc.

Water Communication recognizes that all processes communicating on water have set up specific languages, discourses, lexicography, media planning and campaigns, practices, knowledge transfer or education and engagement.

1.3 HERITAGE AND INFLUENCES OF COMMUNICATION SCIENCES

Water is considered as a "transversal resource". Transversal approaches in Mathematics signify the intersection of spaces or maps. Water's transversal aspect suggests managing water resources with the support of several entities. For example, at the national level, water resources are usually under the responsibilities of the ministries of environment, agriculture, health, energy, education, etc. The difficulty is the coordination of ministries, different authorities or funding programs involved in water management. Similarly to the need for the transversal water resources management, water and communication is subjected to many influences. Many topics, organisations and interactions are building discourses, codes and references on Water Communication.

1.3.1 The Heritage of Communication on the Environment, Sustainability and Resources

If "Water Communication" as a specific expression is not yet fully widespread, research dedicated to Environmental Communication is widely accepted. Research considers separately Environmental Communication, Sustainable Development communication and communication on other natural resources.

1.3.1.1 The Key Influence of Environmental Communication

Regarding environmental topics, Boykoff and Roberts (2007) observed the parallel development of media with the raise of the visibility of climate change. They pointed out the influence of the media in reaching the public attention about climate. They particularly analysed the impacts of such a relationships underlining both the importance of technics and of reaching households with awareness of how human activities impact the environment. In this sense, the similar evolution of communication technologies and the public's uses of those technologies increased the public awareness of environmental causes, discourses and rhetoric.

Traditionally, people consider the first pictures of the Earth taken by Apollo 17 in 1972 as the origins of Environmental Communication. They initiated an ecological awareness impacting people's engagement towards the environment. They developed images and discourses on the preservation of the Earth.

Environmental Communication can be considered as both a field and a phenomenon. Environmental Communication covers all environmental affairs and resources. In this frame, water is included as one of the chapter, or sub-topic under the spectrum of Environmental Communication. It is interesting to remember the transversal characteristics and human-nature relations of Environmental Communication. Environmental Communication is specific in the approach of individual behaviours. It is generally accepted that Environmental Communication wants to interpret on people's perceptions and ways of life, communication is "not only reflecting but also constructing, producing, and naturalizing particular human relations with the environment" (Environmental Communication Theories, Encyclopaedia of Communication Theory, 345). This performing action on individuals from the environment brings symbolic values and actions, unconscious and non-rational behaviours towards the environment.

If looking at the history of Environmental Communication, the field has emerged during the late 70s - early 1980s. Bourg (2001) showed that the ecological awareness originated from the work of George Perkins Marsh based on his book, *Man and Nature: Or Physical Geography* as modified by Human Action. The UN conferences on environment, several essays and publications by journalists (in particular, Rachel Carson, *The Silent Spring* in 1962) opened the environmental cause to public awareness. *The Meadows Report* in 1972 also played an important role in resembling policy makers and raising international visibility and debate on environment. The organization of the first Earth Summit in Stockholm and the creation of the United Nations Environment Programme (UNEP) participated to institutionalize debates on the preservation of the planet. More than key books and events, Bourg (2001) points out the heritage of Environmental Communication from ecological awareness which can be found in ancient texts such as *Critias*, by Plato.

Regarding the setting of scientific research and groups, some first references and research groups are to be found in Norway or the US. In Norway, the philosopher Arne Naess initiated the phrase "deep ecology" in 1973. Her article on "The Shallow and the Deep, Long-Range Ecology Movement" started movements to defend ecological protection worldwide. The movement, despite the controversies and conceptual questions it raised, initiated conceptualization and studies on communicating about human and animals species to advocate for their protection. In the US, Oravec (1984) is often quoted as an important contributor with her publication analysing controversy over a dam in a natural site.

The primary work and its exposition to the public showed the importance of environmental rhetoric and discourse. This aspect represents an important part of current research to address ways companies, policy makers, international organisations frame environment in their publications, discourses, etc. With the acceleration of those studies, research on media covering environmental stories and environmental journalism has been largely disseminated worldwide. Cox (2010) underlines that those two fields are part of the 7 key research areas of environmental adding:

1. Rhetoric and discourse
2. Media and environmental journalism
3. Public participation in environmental decision making
4. Social marketing and advocacy campaign
5. Environmental collaboration and conflict resolution
6. Risk communication
7. Representations of Nature in popular culture and green marketing

Milstein (1993) reminds us that Environmental Communication also includes relations to engagement. The particularity of the environmental cause facilitated engagement and activism from many individuals and organization for the protection of the environment. A large number of studies are dedicated to observing the engagement, lobby and advocacy of environmental discourses and rhetoric. Milstein reminds that Environmental Communication is a "crisis communication" pointing out that individuals will act and react only when supporting directly and indirectly the impacts of changes on natural resources. As such, the invisibility of environmental impacts and the difficulties to prove them scientifically explain why environment communication had challenges in making behavioural changes. Bourg (1990) also shows that the human-nature relationship is related to the human interactions with technologies and will to control the planet.

In this frame, Environmental Communication demands a radical shift that Sustainable Development brought in more concrete words. The official statement and repetition of the definition of Sustainable Development since 1987 is a way to make people reconsider their relations to nature and development models. Bourg (1992) showed that the evolution of wording is revealing of people's non-rational relationships with the environment. From nature to ecology, environment and Sustainable Development, significations are brought to individuals in order to make them aware of their impacts on nature. Recent terms such as "green economy" and "collaborative or sharing economy" are new economic models that are attempting to rethink human-nature relationships as well as model of development in particular for industrialised societies. In terms of definition, Environmental Communication refers to "the planned and strategic use of communication processes and media products to support effective policy making, public participation and project implementation geared towards environmental sustainability" (OECD, 1999).

Regarding the notion of sustainability and environment, Environmental Communication and Sustainable Development Communications are complementary to the building of Water Communication science.

1.3.1.2 Influence from Sustainable Development Communication

The number of initiatives about Sustainable Development[5] and the number of research groups dedicated to Sustainable Development communication confirm its importance and uniqueness. In short, Willner (2006) analysed the specificity of Sustainable Development communication on the account of the GTZ[6]. She identified 4 characteristics of the specificity of Sustainable Development including (1) its complexity; (2) the comprehension gap between experts and the public; (3) the personal impacts on people with "non-rational, e.g. emotional and spiritual dimensions", and (4) risks. This brochure presenting "a conceptual overview" is interesting in summarizing the main characteristics of the discourse and communication pillars for Sustainable Development. In this frame, the Sustainable Development communication traditional curriculum of scientific investigation will include (Willner, 2006)

1. Development and Environmental Communication
2. Social marketing
3. Non-formal and environmental education
4. Civil society mobilization
5. Conflict management and negotiation

The five sub fields are interlinked; they aim to embrace all these aspects for "strategic communication for Sustainable Development". According to Willner, "these five branches of strategic communication interventions should best accommodate the three pillars of Sustainable Development: ecological (particularly branches 1, 3, 5), economic (particularly branches 2, 5) and social issues (particularly branches 1, 4, 5)", (Willner, 2006: 26)

If this definition is engaging in order to understand what Sustainable Development communications cover, it is revealing of the ambition and faith put in communication strategies. Each field represents an important amount of information and research that might be difficult to synthetize and properly integrate to strategic planning. It places much hope and pressure on communication managers and strategies.

Jollivet (2001) stated that the discursive particularity and success of Sustainable Development communication is based on three key pillars: Sustainable Development is a promise; secondly, this promise is for a better world and thirdly, this promise is based on the criticism of the industrialised and frivolous modern societies. The communication of Sustainable Development is different due to this dual capacity to make people reconsider the current state of society and to trigger forces for new social values and rules. Another interesting feature is the relation to time. Godard (in Jollivet, 2001) reminded that communicating about Sustainable Development conveys messages to current generations for future generations. The relation of time will also be present in discourse and rhetoric related to Sustainable Development but will influence public perceptions as well as their capacity to change or react to messages and campaigns.

Since Sustainable Development is also a shifting approach towards traditional economic theories, many are objecting to its communicational potential and "plasticity" (Aim, 2004). Sustainable Development allows an "empty rhetoric" (Godard in Ducroux, 2002) that want to change the whole society and all individuals. In his essay about green communication, Libaert (2010) demonstrated this assumption by analysing the apparition of communication materials and formats since 1992. He described five models of interactions that helped the dissemination of Sustainable Development in the public sphere. Based on the theory of stakeholders (Freemand and Phillips, 2002), Sustainable Development provided the opportunity to mobilize the public on new perspectives, new vision, "re-enchanting" the world without concretely changing its model. Sustainable Development is "the invention of communication" that calls for

---

[5]See Sustainable Development Communications Network, CSD Initiative, Futerra, IISD, ISD, etc.

[6]German cooperation Agency for Development, changed to GIZ (Deutsche Gesellschaft für Internationale Zusammenarbeit) in 2011.

reconsidering communication processes and transformations. His theory points the need of coordinating communication discourses and initiatives on Sustainable Development but more largely, on environment and natural resources. Libaert recommends five pillars to Sustainable Development Communications:

1. The principle of responsibility
2. Modesty and simplicity
3. Holistic communication
4. Participative communication
5. Slow communication
6. Transparency based on the consultation of stakeholders
7. Moderation by controlling the number of materials

This ambitious thesis is nonetheless, a key contribution to the particularity of communication on Sustainable Development. Reaching out the usual criticisms of the communicational potential of Sustainable Development and dangers of green-washing, Libaert initiated a new type of Sustainable Development and Environmental Communications based on "selective communication". As natural resources are scarce and vulnerable, to communicate on natural resources supposes to respect their inner qualities and characteristics, e.g. to limit communication campaigns and to consider them as rare.

Now regarding Sustainable Development communications and its relationships with water, an interesting definition has been provided by the conference in Zaragoza (2008) "promoting the sustainable use of water resources to satisfy the needs of the present generation, without putting in danger the supply for future ones". This statement is similar to the definition of Sustainable Development (Brundtland Report, 1987).

Four main common traits (Hervé-Bazin, 2012) can be identified between Sustainable Development and water discourses:

1. The need to engage the public to preserve a natural resource; this engagement is necessarily global but based on local actions.
2. Because water and Sustainable Development are transversal, to embody its cause is complex particularly at the human and local scale (Godard in Jollivet, 2001; Bourg, 2003; Bourg *et al.*, 2006)
3. This barrier supposes to bring visibility to individuals. Virtual water appears to be a concept attempting to overcome those characteristics (Aubry, 2007)
4. The cause supposes to change of social model including a paradigm shift (Pascual-Espuny, 2007) by promoting responsibility and new ethics towards individuals (Jonas, 1990; Bourg, 2003). The remaining barrier will usually be the non-rational barrier from individuals' perceptions and behaviours as well as their inclination to act (Godard in Jollivet, 2001).

Those similarities between Sustainable Development discourses and the characteristics of water communications show the heritage from environmental and Sustainable Development communications on the codes and rhetoric about water. The question would be to examine its special features as well as its potential links and interrelations with other natural resources.

1.3.1.3 Other Natural Resources Communications

Among natural resources, the most developed area is "Climate Change Communication". If the topic is more topical, it starts to gather an ensuing number of specialists and research centres (for example, Centre for Climate Change Communication[7]; The Yale Project on Climate Change Communication[8];

[7]www.climatechangecommunication.org

[8]www.environment.yale.edu/climate-communication

the Centre for Research on Environmental Decisions at Columbia University[9], Climate Change George Mason, etc.) Talking Climate, a UK initiative based on the partnership of several UK research networks and universities[10], listed a series of research topics and references on their website (Box 1.2). This list with topics in italics shall serve as a reference to identify research areas applied to Water Communication.

Box 1.2 List of terms and research reference for climate change communication, by Talking Climate, www.talkingclimate.org

- Communicating climate change
- Communicating climate science
- Encouraging sustainable behaviour
- Visual communication of climate change
- Making climate science simple & understandable
- Communicating uncertainty in climate science
- Why are people still sceptical about climate change?
- Social norms & social networks
- Using scare tactics: does it work?
- Resources for communicating climate change
- Breaking bad habits & creating good ones
- How to go beyond social marketing
- Language: words & phrases
- Values & frames
- Uncertainty & the IPCC
- Public perceptions of climate change
- How is the UK government promoting sustainable behaviour?
- Climate change scepticism and the media

For other natural resources, when looking at the four elements (Water, Fire, Earth and Air) no similar use of terms and strong research development can be found. As stated earlier, under the "water" definition and scope, the sea is an important water resource of the Globe. Communication about the sea will include various important aspects such as fish and fisheries, coastal zones, marine biodiversity, etc. Today, it is recognized that better communication on challenges related to preserving the sea and its resources, is needed as much as research on communicating about water, fresh water, drinking water, etc.

For "fire", the terms "Fire Communication" will directly include technologies to prevent fire expansion and materials for humans to use in case of fire. As an expansion to fire, it is rare to find such association for energy such as "energy communication", but some emerged on "Renewable Energy Communication" or "Sustainable Energy Communication".

[9]Published guide on the psychology of climate change communication, CRED Guide, 2009.

[10]http://talkingclimate.org/about/ "Talking Climate is a UK-based partnership between the Climate Outreach and Information Network (COIN), the Public Interest Research Centre (PIRC), the Understanding Risk group at Cardiff University and the 'Climate change as complex social issue' research group at the School of Sociology and Social Policy, Nottingham University."

Under the term "Earth communication", research will usually be found for improving communication about preserving the Earth, and be related to the environment or Sustainable Development indicating the large scope of Environmental Communication. If looking at wood, the concept of "Wood Communication" is absent but can be found under "Forest Communication". For instance, the European Commission developed an initiative for the International Year of Forest including the identification of gaps and needs, perceptions and actions to be taken (http://ec.europa.eu/agriculture/). "Air Communication" will only be found to refer to communication technologies for the air sector and is absent to consider specific types of communication processes about the air. About air, research can be found under communication on "air pollution" or "air quality". Most of the research activities will aim at communicating to the public on the health impacts of air pollution or/and air quality.

For other traditional fields of environmental resources, much more materials and references can be found on "Waste Communication". At the European level, many public campaigns about waste have been conducted over the last two decades. It indicates the specificity of codes and processes to address the polluting activities by human beings. If looking at other natural resources including non-renewable natural resources such as heavy metals, examining the codes of the industry of gold, diamond and any other mining industries, is included in marketing communication. In particular, much material can be found in research about luxurious industries, branding and social representations. As such, the idea of limited resources is usually absent except to convey the idea of rarity. In this sense, the common association with water being the "blue gold" points out that research on Water Communication could find further shared characteristics and interesting comparisons with the luxury industry.

On the theme of biodiversity, since the topic is much more identified and historically implemented, materials about "Biodiversity Communication" such as guides and toolkits[11] are numerous, as well as branding publications[12]. Several research centres are dedicated to provide scientific revision of the public perceptions and representations on biodiversity, communication and advocacy strategies, etc.

The outlook on natural resources today signifies common resources and common good. The common good is to the advantage or benefit of all people in a society or group. The notion is related to ethics. Plato, Aristotle and Cicero discussed the notion of common good in ancient Greek times. The notion is mostly implying that people should have equal access to social systems, institutions, or environments. When discussing natural resources, the notion implies that people have equal access to vital natural resources. Since water is a vital resource – with air, the concept of common good is very important in the building of discourses about water. The discourses about water as a "common good", a "universal patrimony", a "shared resource" refer to the notion of common good. It involved responsibility, ethics and equal access. For many, it suggests free access or affordable access to water resources. The notion of common good for water is very important in the discourses of altermondialists and militants. It is also very important in institutional discourses such as the UN, governments or the private sector. Discourses are built on the guarantee of equal access to water services and the protection of water resources for the common good and humanity.

The concluding relates to communicating about natural resources, would be living animals. The term of "Animal Communication" could not be applied since it already refers to an important spectrum of research work on animal languages and transfers of information. The field of communication about animals is characterized by many communication research activities in particular, discourses and advocacy campaigns for animals rights, representations and public perceptions of animals, journalism and media coverage of animals, and branding for animals consumptions, animals social marketing, etc.

---

[11] Communication, Education and Public Awareness (CEPA), UN Decade for Sustainable Development, Scotland's Biodiversity Communications toolkit, Life + European Program, etc.

[12] Futerra Guide, Royal Society for the Protection of Birds, the UNEP, etc.

By shortly comparing with other natural resources wording and categories, many similarities can be found among different themes mainly under the scope of "Environmental Communication". The different terms of the category "natural resources" show the circularity and transversal characteristics with other codes, discourses and identities. The use of a specific term such as "Climate Change Communication" or "Renewable Energy Communication" underlines the interest by researchers, funding actors, or policy makers as well as the need to raise awareness of the public about specific aspects of those issues. The emergence of terms in order to name and frame these ideas underlines their importance in the public sphere. Existing research using such dynamics also reveals the recognition of communication processes holding a certain prestige. Water Communication is not an exception; many other fields influenced the construction of its uniqueness.

1.3.2 Other Main Heritages from Communication Sciences

When looking at a transversal resource such as water, many other topics will influence the building of communication on water issues including religions and culture, health, development models, risk, legal aspects, public service, science and cities.

1.3.2.1 Religious and Symbols: Water Social and Cultural Representations

Historically, religious and symbolic beliefs clearly influenced the perceived characteristics of water and conveyed attributes to impact social representations of water. Many writers and researchers have shown the wide range of cultural representations of water. In short, there are five major symbolic values related to water (Durand and Encyclopaedia Universalis, 1993): (1) life, fecundity and seed growth; (2) medical and health; (3) purifying and rituals; (4) diluvia and ambivalence; and (5) civilization and human utilisations. Taking into account those five categories will help to consider the influence of symbols on social representations of water.

Many researchers have established the links between cultural representations and behaviours in particular; ecological behaviours are interconnected with the personal relationships of individuals with natural resources (Moser *et al.*, 2005). According to Moser et al., the representations of water in terms of immutable resource and quality and quantity will influence people's actions towards protecting resources. De Vanssay (2003) showed that social representations of water are a synthesis of several experiences and might have stronger influence on people's capacity to change or their interpretation of environmental issues and challenges.

These social representations need to be involved with playing an important role in public perceptions of messages, discourses, campaigns, etc. For water, the dimension of locality and territory (Tröger, 2010) It is particularly important considering special relationships with their water nearby (rivers, lakes, sea) and their personal use of water (domestic consumption, hygiene, etc.) At the individual level and its interpretations of water access, further research and elements of interpretations can be found in Water Anthropology.

Water Anthropology will usually cover the different uses of water throughout times and in different societies. In this frame, the anthropology of water is particularly focussed on generating studies of symbolism and its origins. The anthropology of water traditionally contributes to point out the perceptions building social practices and social representations. The interactions between water and individuals as a conceptual approach to individual mind-setting and behavioural intentions are clearly participating in the circulation of conceptions impacting water management, water preservation and water discourses. Brelet (2012) eventually showed anthropological interactions with the history of the UN and water as a guiding resource that contributed to promote the right for water and sanitation.

The nature of water itself and its impacts on people is further analysed by Strang (2004) who values the relationships between humans and nature to explain people's reactions to water issues. For her, biology and physiology are factors influencing cognitive reactions and behavioural intentions and acts. In this frame, she highlights gender issues and impacts on water conceptions quoting the 'archetypal water carriers' being women. The transfer from women carrier for family uses to male water carriers in cities and the invention of pump water changed interactions with water as a resource. The technology changed water interaction into a commercial interaction (Kinnersley quoted by Strang, 2004). Strang further analysed the impacts of infrastructures in the individual conceptions of nature quoting the example of dam as proper example of engagement to preserve water. All these material changes influenced meanings, involving both the technological and cultural, directly impacting social and environmental relationships. The work of Strang contributed to point out the specificities of human relationships with water resources also generated by the anthropology of water.

Anthropological contributions to water studies underlined the relationships between the development of water infrastructures, their legal requirements, water politics and social interactions, social uses and relationships to power (Merill Singer *et al.*, 2010). Merill Singer interestingly quotes the impacts of the "commodization" of water in the case of the Coca-Cola Company and its relationship with the engagement of individuals (Aiyer, 2007 quoted by Merill Singer, 2010). She notably recalled the long-lasting lack of link between access to potable water and health improvements and conditions (Whiteford and Cortez-Lara, 2005 quoted by Merill Singer, 2010). As such the medical anthropology of water and health is a recent research field.

1.3.2.2 Health Communication: From Public to Personal Perceptions

Health Communication is a transversal field that regroups fields and research that aims to enhance individual and public heath. Health Communication aims particularly at changing people's behaviours focusing on information content and mediations; the influence of information on individual reactions (rejection, fear, acceptance, etc.); investigating impacts of public health policy campaigns and behavioural changes. Health Communication has been developed to accompany public communication serving public concerns and challenges. But if Health Communication is usually tied to public communication, nonetheless also other types of organisations are setting strategic communication campaigns and materials (food and beverages industries, pharmaceutical companies, sports, etc.)

The discursive components of Health Communication are ethics and the respect of individual choices; to avoid stigmatization; to consider unequal access to information and social origins; to bring solutions to potential health risks; to accompany long term processes (alcohol, smoking). Applied to water and health communication, we identified five recurrent thematic categories.

1.3.2.2.1 The importance of water quality and combat against waterborne diseases

This field strives to guarantee public hygiene, to inform on drinking and domestic water for personal uses, to inform and to prevent diseases related to polluted water, to educate on links between water and diseases. The historical conquest of water to public hygiene, thanks to the building of infrastructures and the installation of taps in homes, has not been well communicated or informed (Goubert, 1986; Strang, 2004). As a result, the public is still very much concerned about drinking water quality in particular, tap water; scientific research on public perceptions of tap water quality, water resources in their environment or bottled water is abundant.

According to Falconer et al., the public is more and more aware of risks related to drinking water, this implies that communication related to health and drinking water needs to evolve towards a more participative dynamic of informing and designing campaigns. "The judgement of safety—or what is an acceptable level of risk in particular circumstances—is a matter in which society as a whole has a role to play" (Falconer *et al.*, 1999). The communication of water- and health-related issues implies be based on the acceptability of water standards, their understanding and perceptions.

The fear of polluted water both for drinking and its environmental and personal uses encourages negative perceptions of drinking water and potential risk for human consumption, its requires knowing the audience and specific planning (Parkin *et al.*, 2003). Messages from public institutions in order to promote tap water consumption contradict messages from NGOs or global organisations about dangerous water in African countries in industrialized societies. If the geographical distance is understood and campaigns want to create solidarity, the unconscious link with the history of dirty water and non-access to sanitation and public hygiene reminds unclear and may impact people's confusion. They will tend to consume bottled water even if they express their trust in tap water, pay more or use bottled water at home (see Olson, 1999; Ferrier, 2002; Doria, 2006b; Gleick, 2011). In reverse, in countries without generalized water and sanitation access, individuals are constrained to buy bottled water or water from water carrier. They will demand to be able to drink tap water at fountains or from home. Many researchers published about the impacts of tap perception and branding on bottled water consumptions (Wilk, 2006; Talatala, 2008; Ward *et al.*, 2009) in relation with risk (Anadu *et al.*, 2000; Griffin *et al.*, 2000; Doria, 2005; 2006 a and 2009). This observation of various understanding, perceptions and behaviours regarding water and health communication has to be considered with the concept of water security.

1.3.2.2.2 Water security, emergencies and crisis communication

Water security is a recent concept. According to definition by scholars, research institutions and promotion by the UN, water security is defined "as the capacity of a population to safeguard sustainable access to adequate quantities of acceptable quality water for sustaining livelihoods, human well-being, and socio-economic development, for ensuring protection against water-borne pollution and water-related disasters, and for preserving ecosystems in a climate of peace and political stability."[13]

The relationships between water security and health communication implies the need to examine the possibilities for the public to understand potential threats on water quality and water access as well as exceptional events affecting water resources. The screening of such perceived communications should include impacts of such communication on their perceptions and attitudes towards water; information and knowledge needed to promote adequate reactions in such cases. Another potential field would be to look at media interaction with public campaigns, organizational communication, experts' visibility and the public responses in regard to water security and health issues. As a whole, it can further determine the building of social representations on such challenges.

1.3.2.2.3 Drinking habits, nutrition and alimentation

Related to the objectives of enhancing public health attitudes, the promotion of drinking water for better drinking habits and nutrition has become an increasing concern and areas of campaigning over the last decade. Today, if bottled water is usually more trusted than tap water, there is increasing criticism of the quality of bottled water[14] or/and its environmental impacts (Molinaro, 2003; Gleick, 2011). Such criticism has established fears, rejections and new drinking habits impacting people's health.

According to the WHO, the number of obesity cases worldwide has doubled reaching 1.4 billion overweight people, the majority living in industrialized countries. The impact of new living habits including richer alimentation, the consumption of sodas, and less sport resulted in the growth of the phenomenon. For instance, according to the Coca-Cola Company, the consumption of Coca-Cola worldwide has risen from 39 litres per person in 1988 to 85 litres in 2008.

Onufrak *et al.*, (2012) showed that the consumption of soda is related to the perceptions of risk in drinking tap water but also related to other criteria such as region, age, race/ethnicity, income and

[13]Official definition by UN-Water, in UN-Water (2013) Water security and the Global Water Agenda.

[14]For instance, in France, the association 60 millions of consumers published a report shwoing traces of pollutions in bottled water last March 22nd, 2013 provoking a new scandal on tap versus bottled water.

education. Such observations call for more research on the impacts of bottled water marketing and soda marketing on public perceptions and consumptions of water. Research should also consider discourses and the mediatisation of other types of drinking products to create a larger vision on interaction between water, nutrition and public heath. More research could investigate impacts of communication from the food industry on the perception of water resources regarding health concerns and the attitudes of individuals. One practical way of conducting such comparison and studies would be joint research programme with health communication and with marketing and consumers' analysis of their drinking and alimentation regimes.

1.3.2.2.4 Curative values to welfare and beauty

The history of water as being a curative liquid, started in ancient times. In many civilisations, water was recognized for its beneficial properties when used for bathing, thermal cures or as drinking mineral water (see later, the history of Water Communication). With the heritage of "mineral water", the interaction between water and health are constructed on personal relationships with the natural resources. Water is seen as a liquid protecting one's health, general conditions, guaranteeing the protection of the body and appearance. From a health communication perspective, representations and discourses on drinking water are promising for the public. Excluding the aspects of water quality for the health, it is all the "added" characteristics to water that would be interesting to investigate: how mineral or tap water conveys values that will benefit to the individuals and their personal needs. These needs can be for better health but also related to its identity, personality or social acceptance. To look at studies on water marketing and health in extended significations

1.3.2.3 Communication for Development

The concept of Development Communication or Communication for Development is different from Sustainable Development communication. However, one can argue they are related. Sustainable Development Communication is a continuum of Communication for Development. Since it appeared more recently than Communication for Development, the main initial difference is the relation to environment. Communication for Development appeared with the expansion of non-governmental organizations and funding programs promoting models of development throughout the 50s until today. As such, Communication for Development is very much related to the activities of those organizations e.g. the UN organisations and funding agencies from the world, the most influential being the World Bank.

Historically, the Communication for Development (C4D) appeared after the end of the World War II and was initiated by the UN and governments to promote development programs. It included education and dissemination campaigns to accompany the programs at the local scale. The model of communication was based on the sending of information to a passive audience as many considered communication as being a one-way approach. It was also promoting the model of economic development of the Glorious Thirty (UNICEF, 1999). With the economic crisis, raising needs to integrate environmental issues and the recognition of roles played by communities and stakeholders, Communication for Development shifted towards a new approach.

In the sector of water and sanitation, Communication for Development has been very influential to implementation programs. In 1999, the UNICEF defined the Communication for Development model as the basis of the activities and the approach to bring behavioural changes. Throughout the 80s-90s, the activities and programs from the World Bank or the IRC are guided by the approach of Communication for Development. The definition of Sustainable Development in 1987 and the first resolution by the UN in 1991 about Sustainable Development (A/RES/46/167) as a driver for change initiated new perspectives on Communication for Development models. It represented a model to promote technical approach to water and sanitation programs. It contributed to develop discourses and messages about water behaviours and perceptions for more than 30 years, those discourses were based on technological approach rather than cultural and social characteristics.

Another concrete way to understand how much the Communication for Development promoted a model for social and economic development without integrating the population is the evolution of gender and development theories in the water sector. The AFD (Allély *et al.*, 2002) published a key guide summarizing the conceptual approach to development in water and sanitation programs. The concept of equity and the participation of women are recognized during the first Decade for Women (1975-1985). The dissemination of those concepts is accelerated during the 1980s and starts to be implemented during the 1990s.

Since 2006, the World Congress on Communication for Development defined it as a process integrating stakeholders and dialogue. It became ComDev.

> Communication for Development (ComDev) is a social process based on dialogue using a broad range of tools and methods. ComDev is about seeking change at different levels including listening, establishing trust, sharing knowledge and skills, building policies, debating and learning for sustained and meaningful change. (FAO, 2011)

Starting in the late 2000s, Communication for Development became related to Sustainable Development. It aspires to guarantee the protection of the environment and the social participation. As showed the FAO (2011), Communication for Development is an approach that can be applied to many topics: climate change, food security, and water resources management, etc.

1.3.2.4 Risk Communication

Risk communication includes all processes and interactions about risk including risk assessment, natural hazards, risk management, etc. Today, many different discourses on water create tensions, suspicions and fears: water quantity (droughts, floods, scarcity), water quality (as seen in health communication), water management (reliance of infrastructures, water and wastewater plants, etc.), water pricing, water governance (private versus public, corruption, etc.) Many are subjected to controversies. The rising visibility and the naming and framing of risk to humans triggers a scientific and political process of experimentation (Lascoumes, 2002). In this context of uncertainty, words are multiplied and repeated to different public spheres and communication arenas (social networks, media…) concurring to create the "factory at risk" identified by Beck (2001).

Risk communication on water challenges mobilizes a "co-construction" of practices where public participation is integrated to the argumentative and deliberative dimensions of water discourses. They generate interactions between scientific fields, media, politics, the economy, society, cultural and religious belief, or identities. The specificity of the water resource facilitates the multiplication of words and discursive processes. In this frame, risk communication can bring many conceptual perspectives to Water Communication including:

1. Public perceptions of risks related to water quality and quantity, risk biases and beliefs
2. Rhetoric and discourses to explain risks and water issues
3. Controversies on water management risks, their impacts on public perceptions, the role of media, social amplification of risk
4. Communication strategies on risk, risk assessment, situational crisis

1.3.2.5 Right and Legal Aspects, Political Communication

The water sector is intrinsically tied to legislation and entitlement. Brelet (2012) proposes that water guided the history of the UN from its launch to the recognition of the right to water in 2010. From the international to the local level, the number of laws related to water challenges is impressive. Each year, new publications provide insights and updates on water legislation to guide water users and professionals. The communication on the right for water is characterized by several traits:

1. Advocacy for the right for water
2. Practical information and analysis of legal aspects to water management, mediation to accompany its implementation
3. Public and political engagement for water policies

Communication on rights and legal aspects remains one of the most challenging thematic to better water cycle management in the future.

1.3.2.6 Public Communication, Responsible Communication

Public communication is related to all communications with a public service mission. Public institutions produce public communication in the general interest. It includes communication by governments; regional and local state entities; public organisations and enterprises; NGOs. Many public institutions control and play important roles in water management; public communication influences the content of discourses and messages. Many argue that communication about water should engage the public in the name of general interest (AwwaRF & EPA, 2008). The key notion is the importance of responsibility. In this sense, water is also a matter of responsible communication.

Responsible communication applies to wider range of actors since there is no distinction of actors. Private enterprises, citizens, all types of individuals and groups have to communicate with "responsibility". The notion is related to Sustainable Development and sustainable communication. At a practical level, responsible communication assesses all consequences, impacts, risks and opportunities for Sustainable Development. Broadly, the concept of responsible communication involves ethics; it investigates impacts of communication and messages conveyed into societal processes and models. In short, responsible communication is related to Sustainable Development but has a broader impact since it involves ethics and to communicate while taking into account its individual responsibility for the society. In this frame, responsible communication is a key notion to develop campaigns and strategies on water cycle resources and services.

1.3.2.7 Scientific Communication

Scientific communication involved communicating non-technical and scientific topics to non-experts. Water sciences are complex, transversal and human related. Experts and research organisations play an important role in water discourse at the international level. They are also key figures in media or at local scale when setting new projects or addressing the public on water related issues. The specificities of water scientific communication call for specific skills and several terms are used to qualify the transfer of information on water between groups: mediation, knowledge brokering, interfacing, etc. We identified the following fields:

1. Media coverage of scientific expertise, research findings
2. Knowledge brokerage in particular between policy makers and researchers, and vice versa
3. Public perceptions of water scientific communication, public awareness of these challenges
4. Impacts of scientific communication on people's behaviours, strategies and tools to address attitudes

1.3.2.8 Discourses on Cities

In Information and Communication Sciences, many researchers are focusing on the construction of the discourse about cities and the urban life. Cities have their history, scenery, imaginary, and identity. This part is related to communication about territory (territorial communication). This field emphasizes the importance of local references and codes. It considers the building of discourses based on patrimonies, soil, landscape, etc. For centuries, the literature has developed discourses about cities and their way of life. Many are promoting utopic vision of cities.

In the case of water, there are many famous texts about utopic water cities such as Atlantis, the famous lost underwater city. Many famous cities are admired for their relations with water (Venice) or with new architectures on water (The Triton City project). Today, many cities claim to be "City of Water". They seek recognition for the skill in managing water such as Singapore (Newater); Amsterdam (canals and architecture); Malmo (Eco district); or Vienna (water park). The Metropolitan Water Alliance in New York (USA) organizes every year, the City of Water Day since 2007 (www.cityofwaterday.org). Another example, the Victorian Government in Australia organized training to "build the knowledge to create a Water Sensitive City" showing the challenges of water resources management for cities. This example reminds that urban water is one of the major challenges of tomorrow (www.clearwater.asn.au).

Many references in literature or arts put water at the centre of the building of cities. Urban planning and water management are also closely related fields. It is important to consider the relationships between the imaginary of cities and water in social representations as well as in the building of discourses and rhetoric on water challenges in urban water management, the media, and the relationships with social representations of water. The other important field would be the impacts and relationships of water with engagement and citizenship.

1.3.3 In Summary

Water Communication was built through times with the key contribution of many organisations acting at the local level. Those actors are international organisations in particular, several UN organisations; national with both international and local activities such as the IRC; and the far more difficult actors to evaluate: users with their own perceptions and expectations of water cycle management and water services. In this frame, Water Communication is at the centre of many influences: Environmental Communication, Health Communication, Public Communication, Risk Communication, Science Communication and Responsible Communication (Figure 1.6).

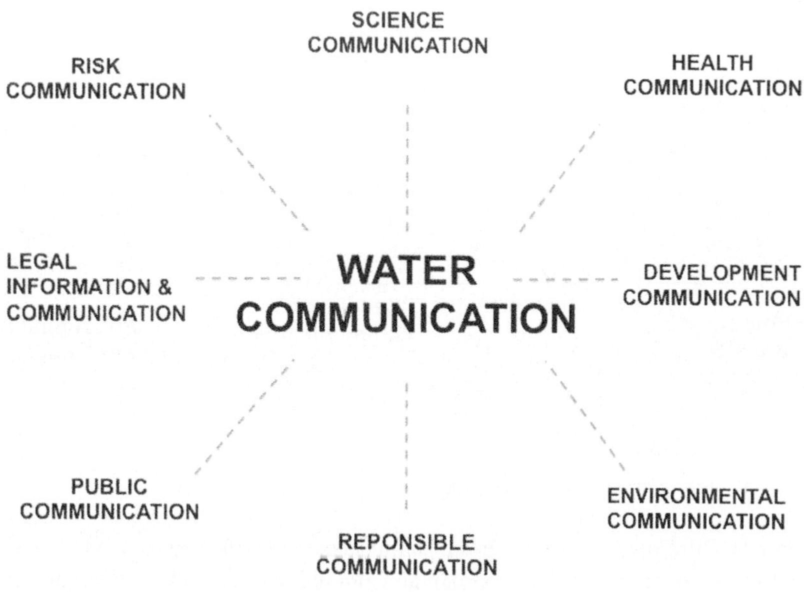

Figure 1.6 Main influences on Water Communication.

The construction of communication about water is influenced by different elements:

- The resource: water. Its cycle, its qualities, its management and services have an impact on the content and topics of communication campaigns
- Three major components influence discourses, codes, messages and strategies:
  1. Religious & personal beliefs
  2. Cultural references about water: myths, uses, and history, etc.
  3. Social codes: water shared behaviours, attitudes and perceptions
- Water Communication is influenced by characteristics from other communication fields (Figure 1.7):
  1. Health communication
  2. Environmental Communication
  3. Communication for Development
  4. Public Communication
  5. Science Communication
  6. Risk Communication
  7. Political and Legal Communication
  8. Responsible Communication

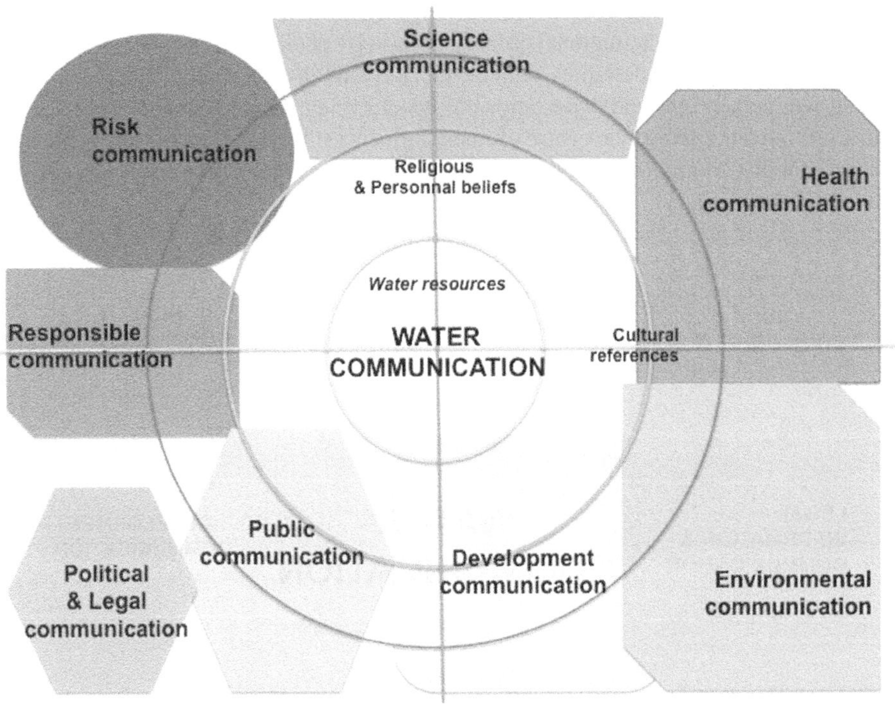

Figure 1.7 The Building of Water Communication.

It is considered that Health Communication has most impact and influence on the content of communication about water, followed, in order, by Environmental Communication, Public Communication, Science Communication, Risk Communication and Responsible Communication. Science and Health

Communication impact directly perceptions on water quality. Environmental Communication and Communication for Development impact the relations to the economic development, the economics of water, the social approach to water services and the management of water resources. They are historically interconnected and Environmental Communication is now influencing Communication for Development particularly since the appearance and development of Sustainable Development Communication. Environmental, public and responsible communication impact directly behaviours and perceptions of water services and infrastructures. Public communication integrates Political and Legal Communication. Political Communication and Public communication affirm the importance of water services and serve the building of general interest. Political-, Legal- and Responsible Communication are interconnected to promote concepts and discourses for the right to water. Risk- and Science Communication impact directly people's capacities to change and relations of fear and doubt towards water management.

2 History of Water Communication

After setting the scope of Water Communication, it is important to consider the history of campaigns and communication on water related issues. The history of Water Communication can provide an overview of the main messages, slogans, logos, and visual identities circulating on water. Communication processes are built through different media, audiences and perceptions over long periods of time. The development of mass media accelerated the development of "Water Communication". The history of Water Communication originated from those developments. Based on this history, the study of current practices in communication will help to identify the key concepts driving the conception of communication campaigns today.

The study of the history of water, its symbolism and most common values showed how water discourses are built. The history water and communication is based on three major types of campaigns and discourses: (1) Thermal cures and mineral waters; (2) Water infrastructures; and (3) Legal systems and Water laws.

2.1 FROM THERMAL CURES TO TAP WATER: HEALTH, SCIENCE AND RISK COMMUNICATION

Historically, when contemplating water, many looked at these religious and social symbols and their impacts on society. Famous civilisations with specific water uses, myths and symbolism transmitted many different values and habits to our modern civilization. Ancient Egypt was the first civilization built around the structuring role of the Nile river. The river facilitated the emergence of a "nation" state conveying the importance of water in the organization of men (Kemp, 1989). The Inca in Peru are known for their advanced irrigation systems and the site of the Machu Picchu still holds its mysteries. Romans are famous for the management of water for their city (aqueduct, waste) and for the collective use of water and bathing with terms. Islam and their conquest of Europe introduced first water tribunals. In medieval times, fountains were synonyms of gathering and had special virtues for fecundity in Europe, etc.

Water is a structuring element in our civilisations (Goubert, 1986). Goubert identified three major steps in the conquest of water that defined the relationships of societies and individuals with water: water as a hygienic resource, water access and the impacts of water access.

In Ancient time until the 18th century, water was pure, a religious value and a privilege. With the French revolution, the privilege of water was abolished (but didn't become a universal right). At this time, the links between water and health were unknown. As shown by Goubert, water slowly became integrated in the matter of public health and related to the fast spread of disease (cholera, typhoid and malaria). The scientific work of Pasteur in the late 1890s and the urban planning by Haussmann in the city of Paris are setting the roots of water management and public hygiene. First standards to guarantee 100 to 120 litres per day per person are initiated, bacteriological analysis of water is launched, and in 1902, the French state votes in a law on public hygiene. The conquest of water is completed around the 1940s-1960s in France. The benefits of the tap at home change the relation towards water. From being pure and rare, water has been synonym of disease over a century and half of conquest to finally reached homes and daily lives. This third stage has deep impacts on individuals mainly in cutting them off from their relationships with nature and water resources.

Pastoureau (1992) mentions that the way water represented in blue is with a "cold colour" indicating that tap water is impersonal. Goubert also states that the arrival of the tap changed perceptions and expectations from individuals towards water. From a symbiotic resource conveying positive and negative attributes of nature – comforting, superior and calling for respect; water became invisible and is now expected to provide new types of virtues mainly in constructing individuals and societal values. From a divine and natural resource, water is considered as a magical liquid for personal uses and the building of one's identity as well as a universal liquid uniting cultures and peoples around on ideal societal vision. Water is a discourse embodying higher values, higher desires and should be perfectly safe for drinking or pure in its representations. This battery of representations is inherited from both communication on thermal cures and mineral waters.

The qualities of "mineral water" were known since the 15th century but without physico-chemical analysis to prove its effectiveness. In France, the state started to rule over mineral water in 1602 thanks to the action of the King Henri IV. Looking at the history of hydrology, the recognition of the medical virtues of water occurred at the end of the 19th century along with the scientific work of Pasteur on waterborne diseases. As a result, mineral waters were first sold in pharmacies. This facillitated two major developments: the multiplication of number of mineral water brands and the enthusiasm for thermal cures (Raynal, 2004).

According to Back *et al.*, (1995), thermal cures originated in Bath (UK). The legend of Prince Bladud initiated one of the first international long histories about the use of warm water. The influence and British colonization in the USA brought the desire for replication. The first thermal leisure spa opened during the late 18th century. The medicinal value of water was accepted and quickly successful. As in Europe, thermal cures gained the USA and developed rapidly. The history of spa therapy (van Tubergen and van der Linden, 2002) worldwide brings additional insights about how water was perceived. The relations to water during its first industrialization are deeply rooted with relation of health, medicinal values, specific places and elitist practices.

The fast expansion of thermal cures built communication processes about thermal stations and first brands of bottled water focusing on health and beauty. Goubert (1986) showed, through the analysis of French journals from the 19th century, that French were very much concerned about beauty and physical appearance when coming to thermal centres. The majority would come for the beauty of their hair, be concerned by their teeth or for their body. When examining the iconography and discourses on thermal cures, the main argument would be about health while promoting a location, leisure time and elitist way of life. In terms of communication processes and social representations, thermal cures or mineral water became directly considered as source of difference, social status. They are also based on mythological references such as eternal youth and eternal beauty (Bachelard, 1948).

In a similar approach, the first bottled waters professed to the qualities and virtues of water for the health. The first bottled waters existed during ancient Egyptian times, they were stored in ceramic pots and know for their values for the health. In ancient Roman times, many well-known waters were distributed as luxury drinks mainly in Europe. The legislation about mineral waters started during the 17th century. France has been leader in developing the market of mineral waters. In France, the concept of "mineral water" originated from King Henry IVth in 1605. A century later, France developed legislation about the exploitation of mineral water; this water was dedicated to royal uses and needs. During the 18th century, the change of political regime brought new legislation. In 1823, a law allowed pharmacists to sell mineral waters (Raynal, 2004). In 1838, Augusta Saturnin Badoit created the first French branded bottled water from Saint-Galmier. In less than fifty years, the development of bottled mineral water is accelerated and many new brands appeared including Vittel, Evian, and Contrex. In 1895, the French catalogue for pharmacists listed more than 80 new brands of bottled water (Figure 2.1).

(a) (b) (c) (d) (e)

Figure 2.1 Posters for thermal stations in France during the late 19th and early 20th century. (a) "La Bourboule": the iconography shows a woman giving water to children. The representation both symbolizes mythology and women's role as a water carrier. In the background, the city is portrayed in a landscape. In the text, "the most reconstituting of mineral water" (*Source*: Wikipedia), (b) "Evian-les-Bains": A woman is throwing some water to drink a glass of water. In the middle, the city on the Evian Lake portraits mountains with the idea of leisure. In the text, Evian water is "the purest of the world, the most limpid and most healthy" (*Source*: www.prodimarques.com), (c) "Salies-les-Bains": A woman with her daughter is sitting in her chair in front of an elegant house and garden of roses. She is reading and resting. In the text, the water guarantees "health thanks to salt" (*Source*: Fonds régional d'acquisition pour les Musées), (d) As opposed to previous one, "Pougues-les-Eaux" shows several people emplacing the social value of going to thermal station. In a quiet garden with the station in the back, people are playing cards or coming back from a tennis game. A child conveys the idea of family gathering. The text announces "a cure of water and rest" (*Source*: Conseil Général de la Nièvre), (e) In a similar setting, "St Honore-les-Bains" represents several people with sports instruments watching at the thermal station from a hilly garden (*Source*: Conseil Général de la Nièvre).

In the USA, the development of bottled water started as early as 1767. In Boston, the Jackson's spa sold the first bottled water to customers. More new brands appeared starting 1800 near the Albany, New York state, Saratoga springs in 1820 or Poland springs in Maine around 1845. The development also started in other areas of the world: in Brazil with São Laurence or in Serbia with the spring Bucolic (www.nmw.co.rs) Looking at the labels of the mineral waters, they argue for their health qualities and virtues (Figure 2.2).

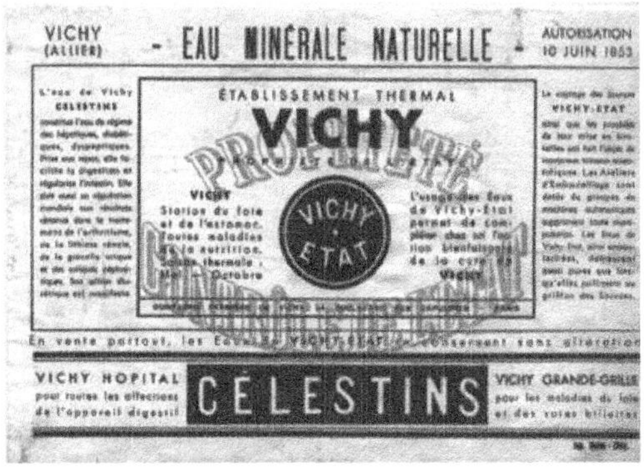

Figure 2.2 Old label conveying medical values. Vichy Hospital, for all digestives affections. Authorized in 1853 and belonging to the State (*Source*: Aquamania www.aquamania.net). The Aquamania website (www.aquamania.net) collected 2,797 stickers from 80 countries.

The mention of medical used is a similar way, as a medical prescription with information about its taste, its colour and odour, and its value for everyday life. The mention of "table water" tends to indicate that the brands want to conquer the habits of customers. Branded waters aim at becoming part of everyday life.

The importance of the origins of the water appears clearly at the very beginning of the development of bottled water, most of them refer to their sources. The location plays a role of guaranteeing the process. It also refers to the place, the natural source, and the spring. In this sense, bottled waters also emphasized their origins since thermal cures are linked to thermal centres. Many brands developed thanks to close SPAs and thermal resorts. They started selling their water to neighbourhood before distributing their water to a wider public. The mention of "mountains", "spring", or "falls" is usually common and largely repeated by the brand.

Those first labels determined the choice and design of labels throughout times. Today, labels on mineral bottled water remain similar to those developed at the end of the 19th century. They provide new types of information since customers are used to bottled water. The information is usually about the mineral contained in the water for transparency and for building trust. Other characteristics are logos, slogans and short texts to convey values and messages about the brand. The labels are serving the general process of branding the water.

In this regard, the progress of branding and first communication campaigns developed new perspectives for selling bottled waters during the early 20th century. They slowly introduced women or men drinking water, slogans or references to codes on thermal cures. Since their health and their bodies interest customers, brands developed new values embodied by their water. They started to build the codes of communication and marketing bottled water. Those codes have been the same over nearly 100 years: health and sports, beauty and youth, environment and patrimony (Box 2.1).

The codes established by bottled water are today, used by public utilities and development programs. For instance, USAID developed a program based on social marketing to promote household water treatment (USAID, 2007). The study underlined the importance of branding with adequate messages, names, packaging. The study particularly showed the importance to adapt to the cultural references and "local sensibilities". The approach was very similar to branding and complaining bottled water since the program addressed several different countries (14 African countries from West, South and East Africa; 5 from Asia and 1 from Caribbean area). The study showed that messages should convey positive values and to be aspirational rather than focusing on quality details and information. The program developed key messages such as "safe water, happy families".

Box 2.1 History of French brands, the introduction of values and codes about bottled water during early mid 20th century

Figure 2.3 Vittel, a joyful curative water (*Source*: Wikipedia).

In 1890, Vittel introduced a poster selling the qualities of its water (Figure 2.3). The poster represented a man pouring water in a glass. While serving water, the man adopted a joyful face while screaming: "Hey! The gout! Pinched! Pushed! Flooded!" "The water of the source must be the table water against arthritis". The man appeared wealthy and is a perfect representation of French bourgeoisie. The visual is easily recognisable. The name of Vittel appeared in upper case lettering and occupied the first quarter of the top of the poster. The brand introduced the idea of happiness, lightness, and good feelings while drinking water.

In the USA, Wyrick Water used the idea of purity and spring water. The label provides indication of the composition of the water (minerals). This label is very common and many brands use the same types of branding their water. The concept is simple: a key message with the name of the brand, the composition of the water, the reference to the origins of the water and the company distributing the bottled water. Vichy also introduced the same types of scenery as used to advertise thermal centres. The poster of the brand projected the image of woman drinking water with the thermal centre in the background. The scene is in frame that refers to painting or elegant mirror. The woman is well-dressed wearing a well decorated hat and beautiful dress. She is sitting in the grass as if she is attending an outdoor picnic. She is holding a glass with a water bottle that looks like a champagne bottle. The mention of "sparkling Vichy" in English is a code for the elite who can speak English. The fact that the water of Vichy contains gas that helps digestion facilitated the message and the image. The water appears to be a luxury and easily accessible since it can be "bought everywhere". This visual is a clear representation of water being a luxury, elitist and accessible to anybody who can buy it.

Similarly, Vichy Celestins recently developed the ideas of health and strength. The slogan of Vichy Celestins mentioned "An iron constitution". The mascot of a knight in

Box 2.1 Continued

his iron helmet and armour reflects the idea of strength and protection. Vichy Celestins is drawing the codes of water as an attribute of health and athletics values. It opened for communication campaigns of bottled water as perfect complement to a great body and for sporting accomplishments.

In another approach, Evian introduced its first baby figure in 1935[1]. Some visuals showed the baby waving at the bottle. In the same frame, the "label" referring to the qualities of the water is a perfect combination of a human character with the informative label. The brand balanced the introduction of a figure with the most common and known endorsement of the bottle. People could easily recognize the brand and identified Evian with the baby. Visuals also showed the character of the mother. Women are holding their babies in their arms. The babies are drinking safe water. The brand also highlighted the particular link between water, health and motherhood.

More famously, over the last 15 years, utilities and entrepreneurs intensively used the marketing methods and codes of bottled water and ancient thermal codes. The over famous war between tap and bottled water has developed new water branding by the government, cities or humanitarian players. Many success stories showed the importance of conveying values on water to gain people's trust and eventually change their attitudes. They all played with the traditional codes of bottled water in order to contradict them (environmental friendly, more economic, local e.g. from the city, citizenship) or to create similar reactions around key values (purity, health/safety, daily use, community).

The legacy of health issues towards water has brought ambivalent attributes to discourses: water can kill and water can bring health. The history of water and health has permitted the rapid expansion of mineral water. The economic competition and the enthusiasm for bottled water in the late 19th explain the development of bottled water. After being sold in other places than pharmacies, bottled water became a daily product guaranteeing safety. Branded mineral water promoted new attributes and values to bottled water such as youth, beauty, health, etc. With the development of safe public infrastructures providing high-quality tap water, utilities started promoting their own water. If the relation to safe drinking water appears to be less predominant today, it still drives decisions and attitudes towards drinking tap or bottled water. It built values and discourses around the chemical and natural state of water as both a natural and drinking resource for humans (Box 2.2).

Box 2.2 Famous initiatives to promote tap water using codes of bottled water Campaigns are positive, promoting values

Utilities

In France, many utilities are famous for their engagement for the promotion of tap water In 2004, the French union for water in Paris' area initiated a key campaign using labels to advocate for tap water (Figure 2.4).

[1]Espace Evian, Exhibition Hall about the history of Evian, Source de Cachat, Evian-les-bains, France.

Box 2.2 Continued

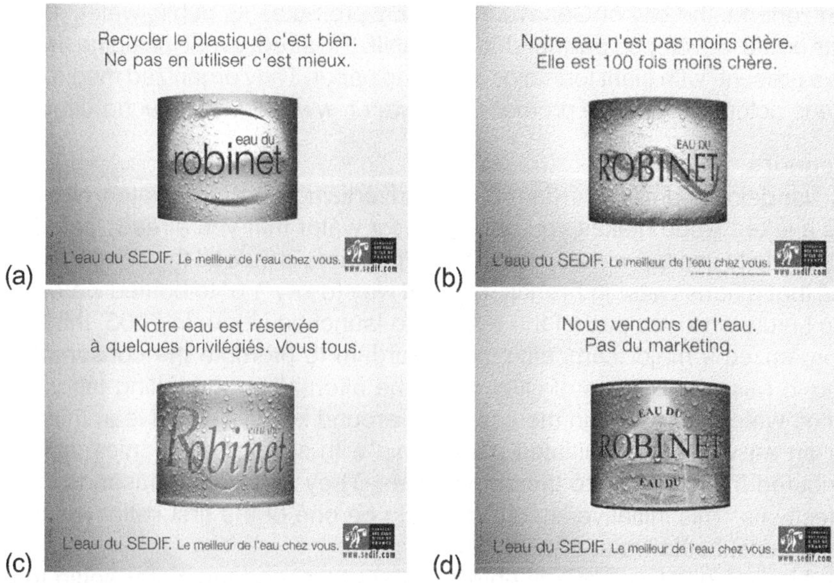

Figure 2.4 SEDIF, the best of water at your home, www.sedif.com - France. (a) "Recycle pastic is good. Not to use it is better, (b) Water is not cheaper. Water is 100 times cheaper, (c) Our water is reserved to few privileged. All of you, (d) We sell water. Not marketing. (*Credits*: Syndicat des Eaux d'Ile-de-France/campagne agence BBDO).

In 2005, the city of Paris distributed for the first time water carafes. Today, each district has its own branded bottle. In 2006, Besancon was the first city to produce and sell its water with the support of a distribution company and supermarkets. In 2010, the city of Mulhouse distributed an artistic water bottle representing one of the most famous building in the city in order to both promote arts and tap water in the region (Figure 2.5).

Figure 2.5 L'eau de Mulhouse, an artistic initiative - France (Drawing by Sarah Stringhetta for Water Stories, 2012).

Box 2.2 Continued

In Switzerland, the city of Geneva intensively promotes its public water. The city has created its bottle, developed partnership with UNICEF, street event including the operation "Water is a present" with giant fountains in crafted paper. They organized media campaigns with famous actors to continue promoting tap water. www.sig-ge.ch/echo-citoyen

Entrepreneurs

In 2003, Vandejong, an Amsterdam-based advertising bureau initiated Neau, a water bottle, as a joke: "Neau makes you pay again for water that you already paid for?" They created a bottle to use tap water in the office. The name is indeed the contradiction of "no" and "eau" that means water in French. It was a way to say "no" to bottled water according to Menno Liauw who was part of the team who launched Neau. In 2005, the advertising bureau organized a major campaign in Amsterdam to promote the consumption of tap water and to raise awareness of citizens on the alternative of drinking tap water rather than bottled water. They design the campaign around key slogans "Neau thirst" or "Take Neau for an answer". They initiated partnership with several companies including Ikea and distributed 5,000 bottles to their employees. They also sold thousands of bottles to several festivals. This initiative is considered to be one of the first initiatives to promote tap water against bottle water (Gleick, 2010).

Since this first initiative, several entrepreneurs launched tap water bottle to promote new ways of drinking water. In the USA, Craig Zucker, a young entrepreneur created a buzz by selling its NY tap water in 2008. The bottle became heavily successful and represented the "water of New York City". The branding played a key role in the success of the bottle. It first played with words as reference to NY's slang and the famous logo "I love NY". The double representation of pipes in building shapes introduced codes of the city famous skyline. Then, messages played with the opposition with bottled water with key phrases such as "No glaciers were harmed making this water" or "Not from the top of some far away mountain". The approach is to against traditional slogans of bottled water.

In France, Gobi has developed a new designed and eco-friendly bottle made in France. Three young entrepreneurs launched their product in 2011 (Figure 2.6). The product established two target groups: enterprises (B to B) and customers (B to C). For the first target, Gobi developed partnerships to replace paper and plastic goblets in companies. The possibility to customize the small piece of paper inside the bottle allows companies to introduce messages and logos. For the second, Gobi designed the bottle for a feminine public based on the design. The hook makes it easy to carry. It relatively small size is easy to put in a bag. Women practicing sports and working women were the main targeted public. www.gobilab.com

Figure 2.6 Gobi, water tap bottle – France (*Source*: Gobi).

Box 2.2 Continued

In the UK, Give me tap has successfully developed a tap water bottle to Londoners in 2011. Edwin Broni-Mensah, a PhD student launched this initiative after missing tap water on his university campus, working with cafes and restaurants. The initiative is widely disseminated through the emblematic figure of the founder. The initiative also dedicates 70% of its revenues to humanitarian cause pleading for the general interest. The slogan "Making our world a fountain" is simple and conveys the values of sharing around the symbolic representation of a fountain. Fountains are public infrastructures. They used to be important point of meeting in Medieval times (Caulier 1990). It gives a sense of citizenship and sharing.

Those both initiatives (Gobi and Give me tap) have developed open-access online maps to help users to find tap water in the cities (restaurants, cafes, fountains, etc.): www.eaupen.net and http://taps.givemetap.co.uk/mindthetap

All those initiatives are playing with the codes of bottled water, health communication and relation to risk. They however add codes from Environmental Communication, sustainability and responsibility. They initiated a war between bottled and tap water.

In March 2010, the *Story of Bottled Water* summarized the conceptual approach to drinking tap water as opposed to Bottled water. It also clearly advocated for new consumption habits and relations to natural resources (www.storyofstuff.org). Later in July 2010, the initiative Bottled Water Matters (www.bottledwatermatters.org) released a similar type of TV spot to contradict the arguments of "The Story of Bottled Water". Both spots summarized the wars over tap versus bottled water.

Many initiatives promoted tap water. For instance, in 2008, during the US Presidential election, Tappening, a non-profit association promoting tap water released a campaign directly aiming at the candidates (Figure 2.7). The "Bottle Stops Here" remembered that bottled water over a nine-month period from 2009 to 2010, the US House of Representatives spent $604,000 for bottled water.

Figure 2.7 Advocacy towards candidates, Tappening addressed Obama and Mc Cain during their campaign for US Presidency – USA (*Source*: Tappening; *Credits*: DiMassimo Goldstein).

The year after, Tappening released another campaign visuals framing false information: "Bottled water causes blindness in puppies"; "Bottled water is the primary cause of restless leg syndrome"; "Bottled water: 98% melted ice caps, 2% polar bear tears" (Figure 2.8).

Box 2.2 Continued

Figure 2.8 The lie, a way to blame the opponent. Tappening claims Bottled Water are lying to customers and uses obvious ridicule statements to show how bottled water can fool customers – USA (*Source*: Tappening; *Credits*: DiMassimo Goldstein).

Their slogan strikes stating, "if bottled water can lie, we can too. There are parts of many examples of lobby against bottled water and many companies took the opportunity to develop new ways of consuming water based on commitment towards the environment. Brita, the water filter, developed an initiative Filter for Good with strong visual ad: people covered with oil after drinking water. The visuals show women and men wearing a white T-Shirt. Their mouth is covered with dark liquid. For the man, the dark oil covers the shape of V-neck T-Shirt looking as his lumbs are filled with oil. The concept was to remind that bottled water is highly polluting, that by drinking bottled water is equivalent to drinking oil. The ad reminds how oil is needed to the process of producing plastic bottles.

In reverse, bottled water replied to those campaigns reminding the quality of bottled water or the economic contribution of bottled water to economic growth and employment. In 2010 the French Union for Bottled Water chose to communicate on the concept of "transparency". It reminds the purity of water and asked the consumer "what do you know about the water your drink?" The campaign plays with the uncertainties related to delivering tap water. The concept of transparency also relates to the context of political polemics over water management by public and private companies.

2.2 FROM INFRASTRUCTURES TO SOCIETIES: ENVIRONMENTAL, DEVELOPMENT & PUBLIC COMMUNICATION

Over the history of civilisation, the management of water resources has been at the centre of social organisations. Many civilizations based their structures on the management of water resources. As stated earlier, examples can be found throughout civilisations. Around the management of water resources, the organization of states emerged. Looking at the state or the city organized the distribution of service. It first initiated the concept of public service and general interest. As such, infrastructures became symbols of the emerging nations. They embody the power and the capacity of the State to organize the society. In the water history, many water infrastructures played a role in the building of civilisations.

In Egypt, the Archimedes' screw, the Nilometer are remaining symbols of ancient Egyptian civilisations. Cisterns played an important role in organizing the society for Mayas (chultuns), in Neolithic villages in Middle East (Syria, Palestine, Lebanon), in Turkey (Stoa Basilica dated before the 5th century before being replaced by the Basilica Cistern (Yerebatan Sarayı) during the 6th century. Qanats in ancient Persia are also interesting water infrastructure concepts for irrigation purposes: a series of well-like vertical shafts, connected by sloping tunnels. Aqueducts are famous for their role in the development of the Roman Empire. The first aqueduct Aqua Appia was built in 312, in less than century, the Roman city built five aqueducts providing approximately 1 cubic meter of water per person, a quantity that many cannot guarantee today.

Fountains played an important role in the building of communities at the level of villages and cities throughout history (Greek empire, Roman empire, Medieval times) and also in three monolithic religions. Several are still today, symbols of their cities: the Alhambra Palace in Granada, the Trevi Fountain in Roma, the fountains of Versailles, the Bethesda fountain in New York, the Peterhof palace in Saint Petersburg, etc. Finally, bridges are also works of art over water. They symbolized the conquest of men and represented key technological achievements throughout History. They usually played an important part in the development of their region and countries.

As fountains are symbols, famous bridges in the world are part of the identity of many cities or between regions: the Brooklyn Bridge (New York, USA), the Golden Gate Bridge (San Francisco, USA), the Great Belt Bridge (Sprogø and Funen, Denmark), the Tower Bridge (London, UK), the Chapel Bridge (Lucerne, Switzerland), the Chengyang Bridge (Ma'an, China), the Rialto Bridge (Venice, Italy), Ponte Vecchio (Florence, Italy), etc.

All those water infrastructures contributed to build the supremacy of Man over Nature. They are the symbols of civilisations, the conquest by men of water, the taming of the environment. The history of the ecological discourse is built on the opposition between man, the nature and technology. Water infrastructures serve civilization to conquer nature and to ensure their economic development. In this context, water is integrated in the Communication for Development and later Sustainable Development. In this context, water infrastructures participated in the emergence of new Nations. They played an important role in building cultural references and collective perceptions of water as a public good, At the national scale, the history of dams particularly shows how dams are important symbols for the building of nations.

The first major dam was probably built in Jordan (Jawa dam) and in Egypt (Sadd-el-Kafara) around 3,000 BC. They consisted of stone walls and bricks. In Egypt, since the dam failed and collapsed after only few years, no new dams were built for centuries. In Mesopotamia, the Nimrod dam was built around 2,000 BC. It was made of soil and helped irrigating crops. The first to build dams with concrete and mortar are the Romans. They also developed first "arch dam" in France at Glanum during the first century BC (Agusta-Boularot, 1997). It represents one of the first achievements of water engineering. The main historical arch dam dated from the Mongol period and is named Kebar. In many regions of the world, dams are built to control rivers. In Persia, several dams were built during the 10th century. In the Netherlands, Amsterdam built its first dam on the river Amstel in the late 12th century. During the 17th century, the building of dams started to become widespread in the USA with the European influence. The artistry of dam from Spain inspired the building of dams in its former colony of California (Yang *et al.*, 1999). In 1832, the French engineer Benoît Fourneyron developed the first water turbine. His discovery initiated the construction of modern large dam infrastructures. The first emblematic large dam of modern time is the Hoover Dam constructed in 1936. Since this date, there have been more than 800,000 dams constructed worldwide.

The building of the Hoover Dam is particularly interesting in terms of communication and discourse. The project was initiated during the early 1920s. After several years of lobby, the project is approved in 1928. At this time, the project was named the "Boulder Dam Canyon". In 1929, six states out of seven approved the project and the preparation work started. The estimated cost was $165 million. In 1931, the construction started and the first concrete was poured in 1933. The dam is achieved in 1935 and a formal dedication ceremony was arranged on September 30, 1935. In his speech, President Franklin D. Roosevelt praised the performance, the achievement and an American new symbol.

> Today marks the official completion and dedication of Boulder Dam. This is an engineering victory of the first order—another great achievement of American resourcefulness, American skill and determination.

The speech is pronounced in a particular context. On October 29, 1929, the USA falls into great economic crisis (Black Tuesday). In 1933, Roosevelt is elected against Hoover, unpopular and seen as the responsible of the economic crisis. The unemployment rate is high and the American society only starts to exit the crisis. The building of the dam has also been highly controversial. More than 100 people died during the construction period due to the work conditions (heat, carbon monoxide poisoning).

The naming of the dam was also disputed. It started to be named by J. Edgar Hoover in 1931 according to the tradition of naming dams after Presidents. The defeat of Hoover initiated a controversy to keep the initial name of Boulder. During several years, maps reflected the confusion since half named the dam "Hoover" and the other half, "Boulder". In 1935, Franklin D. Roosevelt insisted in keeping "Boulder Dam" in his speech and several newspapers covered the choice and described his insistence. After the World War II, the US Congress officially ended a long forgotten controversy since the war and economic depression appeared far in memories. The dam officially became the Hoover Dam. Finally, in 1935, environmental impacts around the construction were disregarded. In less than 6 years after the construction of the dam, no water reached the mouth the Colorado River and several native fish species are listed as endangered today.

From the story of the Hoover dam, several discursive elements appear important:

- The political figure
- The technical performance
- The naming
- The construction process
- The stake of cooperation over sharing waters
- The controversy
- The symbol for the nation

Looking at the recent history of dams, similar discursive characteristics appear. In 1954, Nasser announced his will to build a new Aswan dam changing the name to the "High Dam". Nasser renamed the new infrastructure to distinguish from the first Aswan dam built by the British in 1898. In a context of decolonization and separation from American and UK powers, Nasser had to initiate major initiatives to affirm his lead of the Arab nationalism in the Region. After two years of unsuccessful discussions with the USA and the UK, Nasser announced the nationalization of the Suez Canal creating a political crisis over the Middle East.

Before the start of the construction, the UNESCO prepared to save the treasures of Nubia. Despite the controversy, Nasser pursued his project thanks to the support of the Soviet Union. In 1956, his speech reveals what is at sake for the Nation: his sovereignty.

> They said they will pay toward building the High Dam and then they withdrew their offer and cast doubts on the Egyptian economy. Are we to renounce our sovereign right? Egypt insists her sovereignty must remain intact and refuses to give up any part of that sovereignty for the sake of money. (etc.)
>
> We shall defend our freedom and independence to the last drop of our blood. This is the staunch feeling of every Egyptian. The whole Arab nation will stand by us in our common fight against aggression and domination, etc. (United States. Department of State, 1956)

In his last public speech, he mentioned again the importance of the future dam for Egypt. This speech also underlined how interconnected the two events were. The dam was a symbol against the Occidental world and against the USA capitalism.

> "It is the battlefront on which the Egyptian people and the national army are engaged in the noblest and most violent of conflicts. To the south, on the Nile, there rises the High Dam and the giant power station whose twelfth and last turbine began turning today, thereby signalling the completion of one of the biggest electric power stations in the world."

From the announcement until the opening of the dam, the construction process built the history of one country and its leader. Nasser died before the inauguration of the Aswan dam, the dam was inaugurated with relative indifference. Egypt already benefited from enough electrical power and Saddat was the representative of new hopes and changes for Egypt. In 1971, Saddat pledged the accomplishment of the dam and Nasser's role in its building.

> Brethren, I can only start speaking in this place and on this occasion by mentioning a great man to whom the first and highest credit should go for the achievement of the objective and the fulfilment of the dream. Both Garnal Abdel Nasser and the Aswan High Dam are great symbols. Gamal Abdel Nasser is the symbol of the nation. The High Dam is the symbol of the power of this nation.

Today, the name of "High Dam" is less used than Aswan Dam. The political figure, the building of the nation, the technical performance, the controversy, the construction process, the cooperation: the story of the Aswan Dam regroups all the characteristics of the building of a national discourse for the good of its people and for greatness. Seemingly, the on-going building of the Ethiopian Dam reminds of a similar process.

On April 2nd 2011, Prime Minister Meles Zenawi announced the building of the "Millennium Dam". Few months earlier, the Egyptian Revolution shook the country's political system and many observers saw Ethiopia taking the opportunity to affirm its power in the region. Historically, the waters of the Nile have been divided for the advantage of Egypt thanks to three agreements dated from 1902, 1929 and 1959. In 1959, Egypt was the most populated of the eleven states[2] sharing the Nile. Egypt was a strong leading country in the region. The average annual Nile flow was estimated at 84 billion cubic meters. The 1959 agreement allowed Egypt to use 55.5 billion cubic meters, e.g. 66% of the Nile waters.

Since the mid 2000, the growth of the Ethiopian population changed this geopolitics of the Nile region. In 2013, both countries have a population of 88 million inhabitants. Egypt's population is

[2]White Nile: Rwanda, Burundi, Democratic Republic of Congo, Uganda, Tanzania, Kenya / Blue Nile: Ethiopia, Eritrea, South Soudan, Sudan, Egypt

expected to reach 130 million by 2050 and Ethiopia to count more than 170 million inhabitants by 2050. The change of the political and economic situation advocates for a stronger supremacy of Ethiopia over Egypt in the next few years, etc. The new dam appears to play the role of affirming Ethiopia's supremacy in the region.

> On this occasion, therefore, we, the peoples of Ethiopia, call upon all the peoples of the Nile Basin to turn over to a new page of cooperation and solidarity. As we in Ethiopia build this Millennium Dam with the aim of eradicating poverty, let it be known to all that it will also stand as an expression of our commitment to the benefit of all the countries of the Nile Basin.

The speech of Menes Zenawi pointed out from several angles, the need for such a dam in this region. In a context of international fears over a potential war for water, the leader sent messages to the international community and potential investors to back the project. Since Ethiopia is not as present as Egypt in the circle of international organisations, the country appears to be willing to play a more important political role on the public sphere. They obtained financial support from several countries including China. The government of Ethiopia also needs strong contributions from Ethiopian people and investors in order to finance the infrastructure. The approximate cost is estimated at $4.8 billion.

To formalize its approach, the Ethiopian government launched a website (http://grandmillenniumdam.net/ (accessed 05-09-2013) in order to explain the building of the dam as a strategy of transparency and will for cooperation with all the countries of the Nile Basin. In order to take the lead in the region, the strategy of Ethiopia is to better collaborate with her neighbours, in particular North Sudan.

In April 2011, the dam was renamed "Grand Renaissance Dam" at the start of the construction. This period marks an intensification of discussions with other countries and emerging tensions with Egypt. The building of the dam is also controversial because of the damage it will cause to several minority tribes in the area of the future dam. In August 2012, Prime Minister Zenawi died unexpectedly. The project of the dam became an even more important cause for Ethiopian people. Throughout the country, pictures of the dam with Zenawi's photograph are put in front of public buildings.

In parallel, the media started to cover the story of this future dam. The dam would be completed at the end of 2014; declarations, rumours as well as secret meetings from Egyptian politicians[3] are regularly on the top of news. These reports convey the idea of an upcoming war over the Nile where tensions are already high. In the Blue Nile region in North Sudan, many are fighting over natural resources after the separation into two Sudans in January 2011. The North Sudan recently modernized the Roseires Dam that is located less than 200 kilometres from the future Grand Renaissance Dam.

The story of the new Ethiopian Dam is contributing to the tradition of dams where narrative schemes are similar over times. The idea of taming waters and rivers brings several fears and expectations. Such a discursive construction is found in many other stories of dams. The last famous controversial dams stories included the building of the Three Gorges Dam (Yangtze River, China), Itaipu Dam (Parana River, Brazil – Paraguay – Argentina), Xayaburi Dam (Mekong River, Thailand – Laos – Vietnam - Cambodia), etc.

Dams contribute to the construction of strong political discourse and engagement. Leaders are seizing those infrastructures to affirm their leadership, their figure of power. The dam will both serve the idea of economic development, sovereignty and the idea of independence, engagement and conviction. The controversy over the dam can actually be used in the construction of its own character and refer to one's history as shown by President Lula in 2010.

---

[3]On June 5th, the Middle East Media Research Institute aired a meeting where officials were talking about ways to stop Ethiopia from building their new dam. Official didn't know they were filmed and the news broke in the media two days later.

> When I was their age, I would go to Paraná to protest against the construction of the Itaipu dam, and at that time, people used to say that we couldn't build Itaipu, because Itaipu would be used to flood Argentina. Argentina, back then, got to the point of threatening Brazil with the construction of an atomic bomb, fearing that Itaipu would flood Buenos Aires.

Lula has built his Presidential reputation over his years of engagement in union and his youth as a construction worker. In this speech, he affiliated the construction of the dam with the storytelling he constructed while being the President of Brazil. The example reveals the discursive paradox over time. In 1966, the Brazilian government signed the Iguacu Act[4] for the building of the Ithaipu Dam. Articles 2 and 3 advocated for the interest of the nation, the necessity to exploit the country's resources and the economic potential for Brazil.

Conversley, the non-profit association International Rivers is famous for building an anti-dam discourse to protect minorities' interest and the preservation of the nature. Their research identified similarities in the construction of discourses over dam promises by politicians (Pottinger, 2013).

Water infrastructures contributed to affirm the political will over protecting waters, exploiting natural resources and constructing the national identity. The case of the dams shows the importance of tensions, disputes and crisis over shared waters. Many studies have shown the history of water conflicts and their impacts on the management of resources. In this framing, discourses disseminated shared values, key messages and symbols over the recurrent characteristics promoting water as a public good. Water cities also use this type of discourse to affirm themselves as leader in water cycle management. Those discourses embrace their local communities, their inhabitants and individuals around water resources. Discourses show how stakeholders are answering to the need of people and populations, in particular to their water needs. They built a form of water society.

2.3 FROM WATER RIGHT TO WATER CITIZENSHIP: POLITICAL, LEGAL & RESPONSIBLE COMMUNICATION

Water rights refer to the legislation that organizes the use of water resources from stream to groundwater. Water legislation originated in Roman laws for public use, rejecting the private ownership of water. It is considered as the main influential source of water rights on European traditions over the organization of the legal water systems. The Roman law provided basis to the customary forms of law (Getzler, 2006). There are two main types of rights over water resources: land-based and use-based. The implementation of water rights influenced practices and the organisations of States, but more importantly, the notion of public property, responsibility towards water and water limited uses. It organized the repartition and the use of water contradicting religious beliefs.

In the three monotheistic religions, water is a gift of God, both positive (bringing life, growth, etc.) and negative (flooding, punishment, etc.) values. The laws from God are related to water in many episodes of the Torah or the Bible (Judith, Moses, Samaritan, etc.) In the Koran, the throne of God is on water and Mahmoud receives the words of God covered with water. Islammic society organized the repartition of water resources against this religious backdrop. Water was a free good. Water Rights introduced rules on the acquisition of water and its free use. It introduced the concept that individuals could use water resources for their own needs which is now defined as the "right to water".

The principles of the right to water were defined in 2002. The International Covenant on Economic Social and Cultural Rights defined for the first time on the international level, the founding principles on the right to water.

---

[4]Ata de Iguaçu, June 22, 1966: www.info.lncc.br

> The human right to water entitles everyone to sufficient, safe, acceptable, physically accessible and affordable water for personal and domestic uses. An adequate amount of safe water is necessary to prevent death from dehydration, to reduce the risk of water-related disease and to provide for consumption, cooking, personal and domestic hygienic requirements. General Comment N 15, ECOSOC 2002)

In 2010, the UN and the Human Rights Council recognized water as a human right (UN 64/292 and A/HRC/15/L.14). This historical definition is a result of a long process of acquisition and building of the notion of right to water that can be initiated via water tribunals.

"El Tribunal de las Aguas de la Vega de Valencia" or the Water tribunal was the first legal institution ruling water uses. It was founded around 960 AD during the Arab Umayyad Reign in Spain. According to US historian Thomas Glick, the tribunal probably originated from customary rules of settling water disputes from the Roman era. The tribunal functioned for more than a thousand years. Its main missions were to settle dispute between communities using water for irrigation purposes. Those communities were organized in "huerta" e.g. water users' communities. There are no traces of the ruling since the tribunal was only oral and remaining archives date from the 18th century.

The functioning of the tribunal is particularly revealing of the setting of water disputes. Based on an oral presentation, the users have to defend his cause in front of a series of judges and the community. Users advocate their case following similar procedures to a judicial tribunal. The status of the audience is particularly representative of a public process. The disputes and discussions were concrete times to consider reasonable water uses.

The Water Tribunal acted as a ruling example, a guiding authority and a legal entity. According to Guillem (1975), water tribunals held four characteristics the legal process, the orality, the rapidity, and the economy. In terms of discursive process, they provided a sphere for disputing uses, for organizing a public interest and a shared consensus over a natural resource. They followed the rhetoric of legal institutions and tribunals.

Felstiner (1981) identified the three main steps of raising a dispute in a tribunal as "naming, blaming and claiming". He particularly identified disputes as processes, as "social constructs" (Felstiner, 1981: 631). The emergence of disputes in a tribunal and in the society is equivalent to "studying a social process as it occurs" (Felstiner, 1981: 631). He further argued that judicial disputes and tribunals acted as key element to constructing social sphere, to transform individual grievance into a societal process. In a tribunal occurs a transformation, a social process of change that will act as example for future behaviours and attitudes in the society. Disputes will impact upon the capacity of individuals to respond to trouble, problems, and injustice (Felstiner, 1981: 653). Despite the absence of archives, the water tribunal is known for its customs and its legal organisations. People brought their individual grievance in front of the society for a verdict that will impact the social organisation. The water tribunal was the first place where individual and collective rights over water resources where discussed and finally, divided. As such, the water tribunal of Valence was the first institution to deal with water rights and human right to water.

Discourses from Water Tribunals should be constructed on similar characteristics as legal disputes:

- The need to inform on legal aspects, policies and laws
- The grievance raising disputes between two parties
- The rhetoric of engagement and a cause
- The importance to respect the procedure/contract
- The collective decision or citizens/stakeholders' participation

- The notion of public interest and the process of acquiring such values
- The ethics towards water uses and final decision

In many ways, ancient water tribunals represent extended version of the Greek agora or Roman forum. They are the expression of a citizenship over water resources. At least, they built the first approach to water rights for users and between different users. It constituted the first legal institution on water rights. It was a place to discuss rights and politics about water resources. It played a role in building rights and responsibilities over water resources. Tribunals acted as the building of citizenship and public interest over water resource. The Water Tribunal started to lose its influence during the 20th century but it remains a unique long lasting experiment in the history of human kind. It brought the first conception of rights to water.

The emergence of the concept of "Human Rights" first during the 12th century (Magna Carta) until the first Universal Declaration of Human Rights in 1948 slowly accelerated the promotion of human equity and equal access to the same rights from any individual in the world. At the end of World War II, water rights had been successfully implemented in several countries and among nations over the control of waterways and international rivers (Sironneau, 2002; Smets, 2007; Conseil d'Etat, 2010; see chapter 2). The notion of human rights to water started to develop through international conferences, conventions and the emergence of environmental rights and other human rights such as housing, food, hygiene, etc. (Albuquerque, 2012). The history of water tribunals also participated to the emergence of the public awareness for water resources and public interest.

In 1956, the Interstate River Water Disputes Act in India renewed the concept of disputes. The concept was to organize linguistic disputes between states in order to settle the use, control and distribution of an interstate river. In 1969, the Krishna Water Disputes Tribunal is opened to resolve a dispute between the states of Karnataka, Maharashtra and Andhra Pradesh over sharing of the water of Krishna River, one of the most famous cases of water disputes. The case was first awarded in December 1973 and published in 1976. New hearings and debates were put until a final agreement in 2010 after 43 years of legal disputes, etc.

Looking at the report of the tribunal, much emphasis is put on defining uses: industrial uses, agricultural uses, individual uses. In this context, uses by farmers are balanced with larger uses such as hydropower. The example of the Indian Water Dispute Tribunals clearly put the question of balancing water rights, human rights, the preservation of water resources, and water uses for potential economic growth.

During the 1970s, the advocacy work by non-profit associations as well as the raise of international activities on environmental causes favoured the expansion of the notion of water as a human right. The creation of WaterAid in 1981 at the beginning of the first UN water and sanitation decade initiated the first initiatives to lobby policy makers from the civil society for water and sanitation access. The UK based non-profit association is famous for organizing lobbying actions towards policy makers and to engage citizens for water access and sanitation. In the US, Clean Water Action is created to lobby for the Clean Water Act voted by the US Congress in 1972. One of the missions of the association is to ensure "clean, safe and affordable water" that are the three key principles of the future right to water, etc. In France, Eau & Rivières acted for the preservation of rivers and engaged actions towards the EU legislation. The actions of such non-profit associations influenced the political and international sphere to admit, define and recognize the right to water.

During the 1980s and 90s, more water tribunals were created throughout the world. In the Netherlands, a first International Water Tribunal was organized in Rotterdam in 1982 and followed by another in 1992, in Amsterdam. The goal was to raise awareness on the pollution of the Rhine Rivers. Another Water Tribunal was created in Brazil in Florianopolis in 1993 (Caubet, 2002). The Central American Water Tribunal was created in 1998 (Maganda, 2010). Similarly, their missions were to raise awareness on water pollution and issues related to water resources management. They all have in common to raise awareness on water as a public good, to arbitrate public interest and question individual water rights, collective water rights and human water rights.

During the 1990s, the advocacy lobby for human right to water raised more discussions over management of water resources. First militants raised the issues of efficient water management and oppression of local communities for the interest of economic development. As part of the tradition of water disputes, Vandana Shiva became one of the first voices to lobby for the right for water against the uses of water by large companies, her most famous case being against Coca Cola Company and private companies[5]. Her engagement is related to anti-globalisation movements that rose with the decolonization during the 1950s and the criticism of the capitalism during the 1970s.

A new Water Tribunal appeared in Costa Rica between 2000 and 2003. This initiative aimed at promoting ethics over water resources in a situation of urgency and humanitarian crisis. The question of water uses, right to water and water management in a context of limited resources became a global challenge. The UNESCO published a brochure about "water and ethics" showing the construction of water ethics (Delli Priscoli *et al.*, 2004). Many are questioning how to involve individuals in the promotion to the right to water preserving the general public interest.

Maganda (2010) argues that Water Tribunals are building social participation to water management. They are ways to include the public to discuss water policies. They can bring information and education on legal aspects. They are ways to ensure social participation and to introduce ethics into national and international debates. According to Maganda, Water Tribunals are ways to debate and to mediate over the implementation of human rights and balance the different water uses. The legacy of Water Tribunals brought more involvement from citizens over the right to water.

The media started to cover and framed what is now well known as "water wars" e.g. disputes over shared water resources. The challenge of organising the repartition of shared waters questions water resources management. Worldwide, disputes concern the contractual relationships between public and private companies managing water resources. Different types of public and private partnerships exist but media tend to focalise on cases of failed collaboration in particular, in South America or South Africa, and in recent municipalisation in France.

In France, media focused on the actions and achievements made by French private water companies. Several reporters criticized and blamed private companies showing irregularities and cases of corruption. In France, this dispute is famous between journalists and CEOs of the major private water companies (Lyonnaise des Eaux and Générale des Eaux). In 2001, Gérard Mestrallet replied in a column published a month after 9/11. He mentioned the "real battle for all' calling to focus on developing water access rather than focusing on private and public management. In this frame, several initiatives from non-profit organisations developed the militant discourse for the right to water against private water companies.

In December 2001, several militants from the world including Vandana Shiva, Ricardo Petrella and Danielle Mitterrand created the French Association for the World Contract of Water. In 2003, the US non-profit association initiated a campaign "Water for All, campaigning to keep water as a public trust. They published a brochure based on the recurrent discursive characteristics for the right to water: the claim is to protect minorities from not having water access; the rhetoric is based on the civil engagement and the need for public participation; the values evolve around the public interest, the knowledge of legal aspects and the call to change laws and water contracts; the question of ethics and blaming of the second party e.g. private water services.

---

[5]Ecology and the Politics of Survival: Conflicts Over Natural Resources in India (1991); Water Wars; Privatization, Pollution (2002); Globalization's New Wars: Seed, Water and Life Forms (2005)

In 2005, the City of Varage in France launched the "Call of Varage" with the support of the Foundation France Libertés created by the former French First Lady Danielle Mitterrand. The city called other cities to stop water management by private companies. Danielle Mitterrand already known for her positions in favour of minorities became a strong militant for the "right to water". In 2010, France Libertés launched one of the most famous militant campaigns for the right to water. This campaign started in June 2010, a few months after the municipalization of Parisian water services by the City of Paris and only one month before the vote by the UN recognizing the right to water.

This campaign was based on visuals printed in media and in public place all around France. A short radio spot was also broadcasted in many national radios. The concept of the campaign was built around the imaginary and the origins of "water carriers". Before piped water access, water carriers used to deliver water to inhabitants for their services. They were from popular classes, with low revenues and no education. In the French revolution, archives showed that water carriers were part of numerous popular rebellion and active members of the French Revolution (Goubert, 2002). The campaign built on this imagary. According to the foundation, the name is also from a movement for the right to water originated in Canada and Brazil.

Philippe Starcks, a well-known French designer, initiated the campaign and designed a new water bottle for the non-profit association (Figure 2.9). The innovation was in the shape of the bottle that can only be rested on its side and not on the bottom as usual water bottles. The large rectangular shape of the bottle permits to print message on the bottle. The clothing designer Agnès B. offered to create her own bottle with her signature, it allows for the branding of the bottle as well as the endorsement by other celebrities. Visuals of the campaigns consisted of many famous French celebrities holding the bottles in funny ways or referring to their professions (playing instruments, shooting a movie, etc.)

The campaign has been largely successful and one of the first campaign of the kind in France. Today, the "Water Leaf" is still sold online or distributed during events for public water. The campaign has influenced the establishment of codes, references and discourses for public water management. After this campaign, several initiatives at the European level accelerated the movement:

- The documentary "Water makes money"
- The several petitions against water privatization at the national level in Italy, Spain and Greece
- The organisation of the Alternative World Water Forum in Marseille during the World Water Forum

Last March 2012, the European Federation of Public Service Union initiated an online petition to promote water and sanitation services for all. In less than 1 year, the initiative gathered more than 1 million signatures and reached 1.5 million signatures in the summer of 2013. The campaign is the first European citizen initiative about water management.

The communication campaign was promoted through a simple question: "is water a common good or a commodity?" The rhetoric is based on the similar construction as identified before: the legislative context (the UN Water right 2010, the European petition); the blaming and grievance to guarantee water access to the poor; the engagement of citizens; the collective participation and engagement; the question of ethics and values around the public service and public good. Many ambassadors endorsed the petition to gain more support from the public. The cause is to protect universal human rights. In the design of the communication, several codes are transmitting the value of engagement and fight as well as the importance to the individual of claiming their rights.

The history from water tribunals and development of discourses over the human right to water opened discussions about water uses, ethics and balancing priorities for sharing water resources. The question of values around the use of water brought debates and disputes over values related to the concept of

Figure 2.9 "Water carriers" with the "water leaf" a campaign for the public management of water resources (*Source*: France Libertés; *Credits*: Joëlle Dolé). (a) Agnès B., Fashion Designer (b) Bertrand Delanoë, Mayor of Paris (c) Danielle Miterrand, Former First Lady (d) Philippe Starck, Interior Designer (e) Yann Artus Bertrand, Photographer and film maker.

public interest and the question of values around the concept of "water citizenship". Water Communication exists at the crossroads of legal, political and public communication but fundamentally raises the issues of responsibilities and responsible communication. It contributes to a social process questioning the social contract and the organization of water access.

2.4 IN SUMMARY

The history of Water Communication highlighted the main important characteristics of the building of water discourses. They are built around different values and related to the use of water resources. From water symbolism and religious beliefs, individuals and societies have built symbols, discourses and practices around drinking water, water infrastructures and the water rights. Each discourse embraces different values and unifies the society.

Drinking water preoccupied individuals and communities with its effects on health, hygiene and public health. It has since evolved from being a vector of diseases to an elite drink and is now moving being towards an engaged and environmental friendly way of drinking with the modern promotion of tap water.

Water infrastructures and management served the abilities of human beings to conquer the nature and to develop their economy based on their knowledge and new technologies. It structures the organization of the societies and encouraged discourses promoting shared values around key infrastructures at local, regional, national and international - levels. It brought discourses on existing nationalism over water indicating structuring elements over setting water societies, society building themselves around water resources and water management.

The last pillar is related to the development of water rights. The concept of water rights developed discourses, attitudes and behaviours about water uses of water. It questioned ethics about water resources and participate to the emergence of the right to water that calls for duties and responsibilities. As a public good and a universal good, water suggests establishing responsible actions as world citizens e.g. water citizens.

The construction of those discourses can be summarized into bringing new perceptions, disputes, and values towards water resources management. They are built over the main communication fields influencing the building of Water Communication (Figure 2.10). The specifications remain the relations to the water cycle, water perceptions, religious beliefs, cultural references and social codes. The analysis of the water vocabulary and its visual identities will help to better identify its specific codes over times and in current practices.

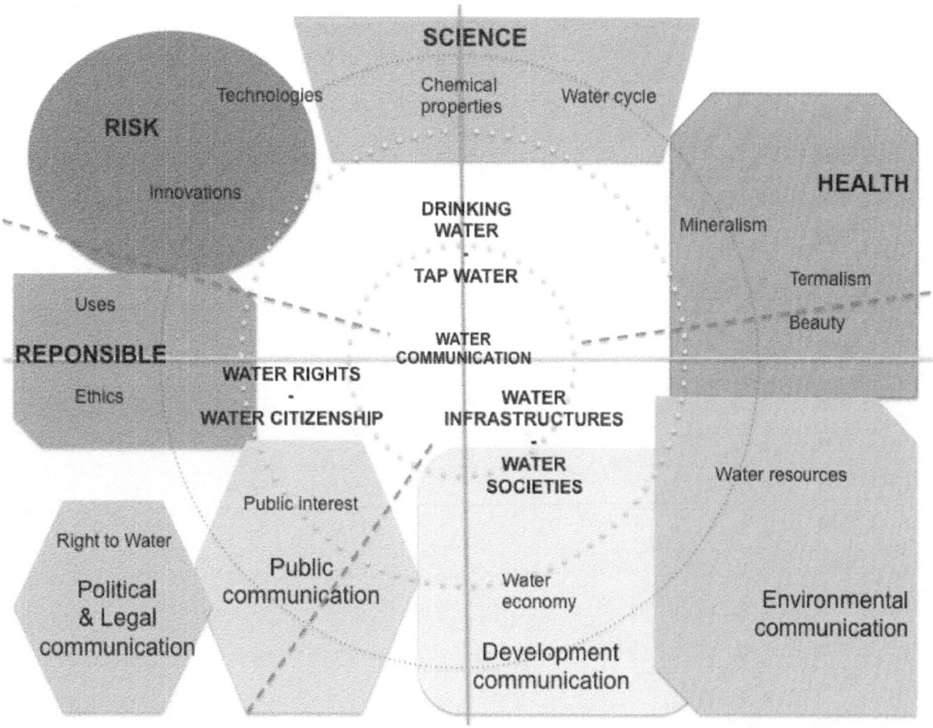

Figure 2.10 History of discourses on water. The construction of discourses is guided by three key principles: drinking water, water infrastructures and water rights. They are interconnected and overlapping.

3 From water epistemology to slogans

From a more practical point of view, we will need to define "Water Communication" taking the same steps and approach offered by research and definitions on "Climate Change Communication" or "Environmental Communication". In order to consider Water Communication as a specific research field, we need to clarify its epistemology, linguistics, visual identity and symbols from the analysis of "water".

3.1 WATER EPISTEMOLOGY

The definition of the word "water" in the epistemology and in the dictionary provides clear understanding of its attributes and characteristics. First, looking at the epistemology, linguists believe the word had two root words for water: *ap- and *wed-. The first is "animate" referring to water as a living force; the second referred to it as an inanimate substance. It brings the symbolic value of water and its vital aspect to human being.

In the encyclopedic dictionary, the word "water" is defined through four key meanings:

1. H2O: Chemistry/Physical characteristics: A liquid of hydrogen H_2O qualifying its chemical qualities. This liquid is also known for its transparency and absence of odours.
2. Natural water: environmental resource: An area of water referring to natural resources and environment. This qualifies water as its natural state bound to the Earth, the different surface waters.
3. Drinking water: human vital need: A body of water adequate for some purpose (as navigation) or human needs framing its link with human uses. Water is understood as a vital resource for men and women.
4. Civilized water: society driver and structure: Waters in plural brings to legislation or to the jurisdiction of a particular country providing water into the context of civilization, its organizing capacities.

The four pillars build the understanding of water from its primary acceptance and pure – chemical formula to the most achieved conception of water as a resource structuring human and social organizations (Goubert, 2002). We first illustrate these four accep tances in a simple graphic representation of water significance and perceptions as shown below. The understanding of "water" evolves from the first meaning (H2O molecule), to the second acceptance as a natural resource, the third definition related to human drinking water and finally, to water as a structuring element for human organization. Further research on perceptions on definitions could be interesting to conduct as well as testing this idea of a cycle of understanding by individuals.

From these four pillars, a series of words, topics and activities can be organized and support the categorization of public perceptions around these definitions, acceptances and social representations based on the lexicography of water wording (Figure 3.1).

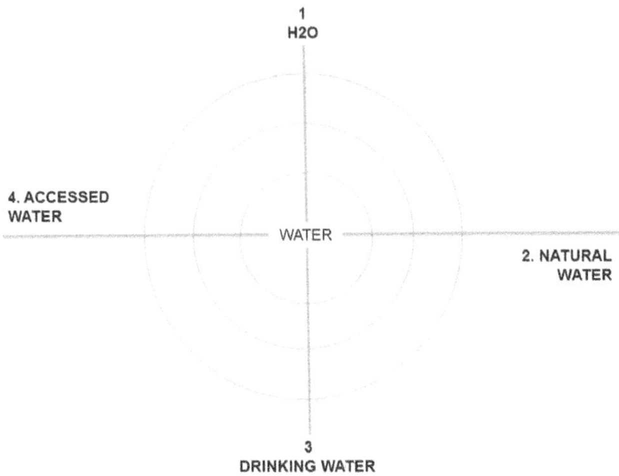

Figure 3.1 The four pillars of the definition of water.

3.2 WATER LINGUISTICS

The following short, non-exhaustive list of terms is classified by categories related to water, the water sector and its different activities and acceptations. This list represents a useful tool for analysis of content when setting a strategic communication campaign in particular, to understand meanings, perceptions and symbolic references related to each word and group of words in the target groups/communities.

Based on this listing and categories, a combined list of words with perceptions and acceptance by the public can establish "four pillars" to understand water significations and epistemology. This combined figure provides a useful classification to understand and to perceive water complexity. Table 3.1 shows a process to understand water as a liquid, a resource, a cycle, a service and its management, and daily use by individuals.

Table 3.1 Short listing of the lexicography of water.

H2O	Natural Water	Drinking Water	Civilized Water
Resource (Chemical)	**Resource (Natural)**	**Resources (Potable)**	**Resource (Accessed)**
Chemical molecule	Stream, Puddle	Surface water	Tap
Chemical formula	Pond, Aquifer	Groundwater	Artificial boring
Two atoms of hydrogen and one atom of oxygen	Lake, Watershed	Mineral water	Drill
	River, Ocean	Spring water	Traditional well
Compound	Sea		Water tank
Universal solvent	Freshwater		Water cistern
Dihydrogen monoxide hoax	Springs		Desalination
	Hot spring Geyser		Irrigation
Heavy water	Glacier		Industrialized water
Light water	Artesian well		Virtual water
	Waterfall, cascade		Sewage

(*Continued*)

Table 3.1 Short listing of the lexicography of water. *(continued)*

H2O	Natural Water	Drinking Water	Civilized Water
State of water	**State of water**	**State of water**	**State of water**
Liquid, Solid Gaseous Snowflakes Crystals Ice Ice cap Ice sheets Water vapour Water drop Dew Steam Cloud Fog, Mist	Snow Rain Rainfall Hail Storm Flooding Meteoric water Mineral water Brackish water	Tap water Bottled water Drinking water Potable water	Reclaimed water Reuse water Purified water Recycled water Highly processed water Used water New water Distilled water Deionized water Purified water Wastewater Screened wastewater Grey water
Technology	**Technology**	**Technology**	**Technology**
Distilled, solvent, reactant, solute, catalyst, etc.	Collected, harvested Hydrometeor	Filtration, boiled, Coagulation Water pasteurization Bottle Ionization, oxidation, chlorination	Membrane filtration Reverse osmosis Ultrasound
Supply/	**Supply**	**Supply**	**Supply**
	Natural pool, channel or waterways	Artificial boring, drill, traditional well, tank, cistern Artificial lake and reservoir Water treatment and Waste water Treatment: primary and secondary treatment methods	Pipes, Waste water Treatment Plant (Tertiary treatment)
Use	**Use**	**Use**	**Use**
Basis of fluids	Drinking water	Societal uses	Facing scarcity
Distribution	**Distribution**	**Distribution**	**Distribution**
Variably present in gaseous, liquid and solid states	Unequally present in the universe and on Earth	Unequally guaranteed by Nations	Rare, depending on water availability and technologies

(Continued)

Table 3.1 Short listing of the lexicography of water. *(continued)*

H2O	Natural Water	Drinking Water	Civilized Water
Characteristics			
Tasteless	Pure	Potable	Transformed
Odourless	Drop/Dew	Washing/Daily use	Industry/Agriculture/City uses, etc.
Colourless	Rituals		
Absence of toxins, pollutants and microbes			
Scientific expertise			
Chemistry	Meteorology	Water quality	Water
Physics	Hydromorphy	Water management	
Biology	Hydrography		
	Hydrogeology		

3.2.1 Chemical Resource

Water is a component from the universe. It holds chemical characteristics. Water is instable, unexpected, with many unknown qualities. Water is a scientific element. It is a field to discover. Water is a liquid, without colour or symbolized in white and blue.

At this stage, water is considered in its chemical properties. The main professional applications for this stage are Chemistry, Physics and Biology. The main representations are the chemical representation of water (H2O). It is considered as a "fluid" or a "molecule".

3.2.2 Natural Resource

Water is part of nature. It holds special qualities that are mysterious and unpredictable. Water is related to the Earth. Water is pure, it symbolizes a certain perfection. It can be long awaited and feared. Water is usually coloured in blue and linked with green, the symbol of the environment.

At this stage, water is considered as a natural resource. Water is part of the Earth; it flows in the environment. The main professional applications are analysing this resource in its natural state. It includes Meteorology, Hydromorphy, Hydrography, Hydrology and other studies of water throughout the Earth e.g. Glaciology, Limnology, Oceanography, Eco hydrology, Hydrogeology, Hydrobiology, etc. Many representations are water in nature, professional applications (water distribution, weather) and water's natural cycle. In this acceptation, water is pure and natural to drink. It holds mysterious powers or qualities. It confers values to communities and to people. As such, natural water resource is related to rituals and religious beliefs. The attributes of most bottled water campaigns and slogans fold into this characterization of water.

3.2.3 Potable Resource

Water is for drinking use. It holds societal attributions and interpretations. Water is a resource for humanity. It is shared but it is also managed to answer to social uses. Water is vital, social and universal. Humans can dominate water resources; but it can be source of inequities in particular, in terms of gender. The unequal geographical distribution of water is source of social and global inequities.

At this stage, water is seen as a potable resource. People will understand that it is a domestic cycle but won't seek understanding or learning about this cycle. The main professional applications are related to water management for primary and secondary treatment (making water potable and wastewater treatment preserving natural resources). Water resources answer to human needs. They are cut off from their natural state.

At this stage, the mysterious meanings or inner characteristics of water as a molecule start to be forgotten and are rarely perceived. Potable water is structuring social organisations and civilisations but is integrated to development. At this stage, water becomes invisible. The process separates pure and natural water with potable water. This distinction and detachment are due to water supply in particular, tap and automatic flows of water in everyday life. At this stage, Humanity is divided and many individuals don't have access to potable water. Main representations of water include water's natural and domestic cycle. Many representations depicted wells, drills, and pumps to embody the lack of water access. In terms of daily use, bathing, tap, drinking water classes are very common. Major water infrastructures are also frequent such as dams. It is interesting to note that they can create many different reactions (pride, polemic, rejections).

3.2.4 Accessed Water

Water is taken for granted. Water is accessed e.g. water is part of everyday life and habits. Water can be accessed any time, anywhere for any types of uses. Water is a technology, a science and a human competence. At this stage, water is part of many complex cycles, uses and management devices. It is possible to produce pure chemical water. Thanks to technologies and innovation, humans hold the ability to reproduce water, its cycle and its resources.

There are many professional applications to develop access to water. They are related to the tertiary water and wastewater treatment. They are also including advanced infrastructures to manage water and wastewater resources. Accessed water is a response to the growing water demand and global development. It is an answer to scarcity and hydro-climatic events. In this acceptation, the public ignores accessed water. The cycle is too complex or can be rejected. People perceived accessed water as artificial, industrialized and technological. In this process, accessed water eventually reaches the chemical stage.

The evolution of wording from the four pillars of water definitions shows the induced complexity by water treatment methods and technologies (Figure 3.2). At the stage of "pure water" or based on its "chemical attributes", few researchers intend to interpret the perceptions of people. The only considerations and analysis will be made based on the cultural, historical, religious and symbolic general understanding of what water might represent in the studied communities or at national scale. Shifting to "natural water" versus "tap water", many book publications intended for the general public have been analysing the contradictions of choosing bottled water or tap water. At last, due to its complexity and status of being "non natural water", recycled water is clearly an advanced field of marketing and public perceptions' analysis. Facing the global water crisis, many entities and research institutes have invested in accompanying the public in understanding water processes and the production of recycled water in particular, in the USA, Australia and Singapore.

This panorama of words is based on the scientific work related to understanding public behaviours and efforts to better target their clients' expectations and needs in terms of water services. These contributions can be found when analysing public perceptions for water quality (Langrod *et al.*, 2000; Pendleton *et al.*, 2001; Doria, 2010; IFOP, 2011; etc.); perceptions of bottled water (Burlingame, 1987; Hurd, 2003; Doria, 2006; etc.); perceptions of reuse water (Bruvold, 1970; etc. 1991; Dolnicar and Hurlimann 2009; etc. 2011; Macpherson *et al.;* 2010; etc.); studies to talk about water (Simpson and Stratton, 2011; Mac Pherson and Slovic, 2011; Howe, 2012); and surveys conducted at the national level. Methodologies based on analysing both wording and understanding of new water provisions methods or technologies are widely used to predict customers/individuals behavioural intentions. Today, the Internet facilitates the work of lexicography and the analysis of words. The most representative device is the "word cloud". It allows picturing the most used words or its frequency. They are more and more developed and integrated in professional practices for communication purposes.

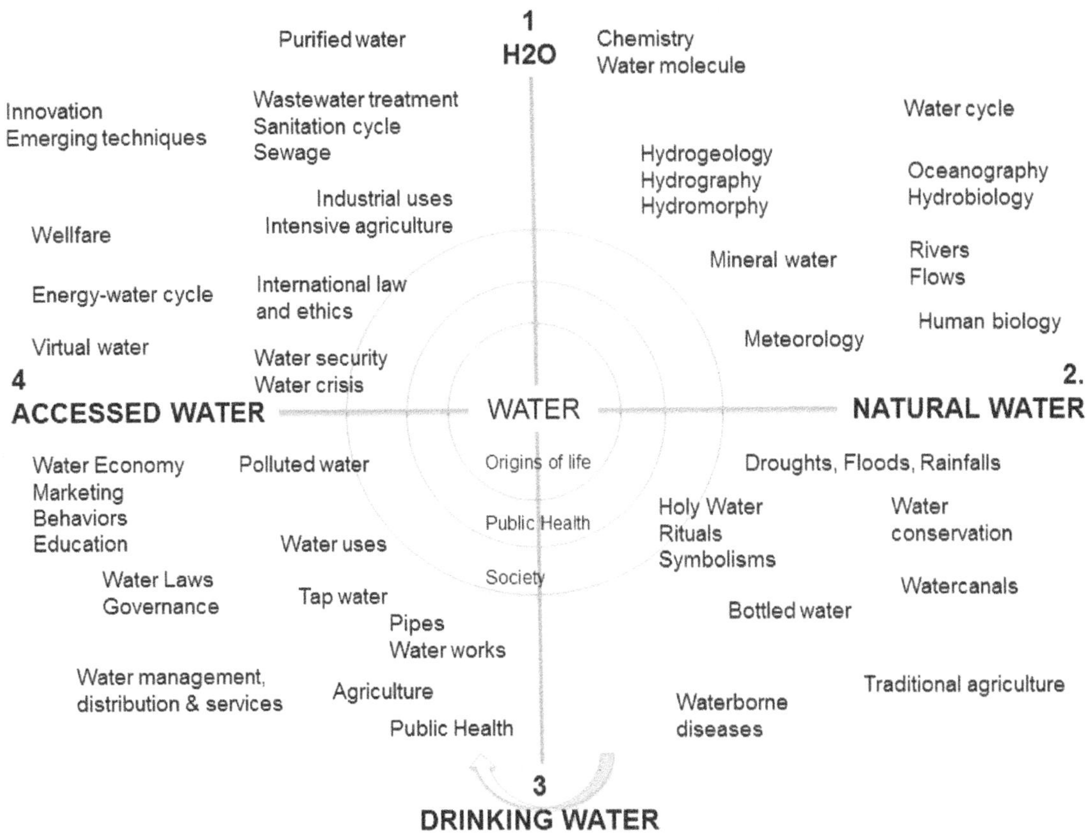

Figure 3.2 The four pillars, detailed lexicography and spectrum.

3.3 WATER VISUAL IDENTITY

Water is an inspiring element to arts for its transparency, its fluidity, its reflection of light (Morelle, 2007). As for symbolism, water is an important subject in arts and iconographic representations. As one of the fourth elements, water is represented throughout seasons, in all different states, in the mythology, in religious icons, in daily life, in landscapes, etc. Water symbols influence the use of logos. Artistic representations and significations will be analysed with specific communication campaigns we used later as example to communication strategies and messages.

3.3.1 Water Symbols

Water is the most abundant chemical formula worldwide. Its first and most famous symbol is H_2O. These three signs are very common and people will usually know and understand the reference to water. The traditional visual representation of the chemical molecule has been an inspiration for many in their logo (Figure 3.3).

Cannes Water Symposium

Figure 3.3 Logo using references to the chemical properties of water. "H2O" is written in specific font and brought into a wave with stars for the Cannes Water Symposium. The Stars are references to Cannes, the city of the famous Festival where as the blue and H2O refers to water.[1]

The mention of H2O is commonly integrated as common knowledge; it plays an important role in conveying the idea of a molecule, a vital element and human value. The symbol is repeated, designed, reviewed as a logo, branding a chemical symbol into an identity with its own codes. The letters and the number allow flexibility in designing that can provide to "H2O", a wide range of significations. Looking on Internet, more than 12,400,000 results appear when searching for logos "H2O". Many refer to consultancies, operations or technologies services. The attributes of the H2O will serve in many contexts, to convey the message.

- Universality: H2O is represented in simple font with no other references. It is usually branded in blue on goblet or carafe.
- Naming: The "H2O" will serve as the name of the organization or the program. In the USA, H2O is a wireless card service. In Australia, H2O is a TV Show about mermaids, etc (Box 3.1).
- Conveying values: The reference to H2O is used to refer to specific values, mostly environmental and educational (Box 3.2).

---

[1]Other examples of this are (a) H2O Africa Foundation - H2O is written in heavy font with the shape of the African continent, a reference the activity area and name of the Foundation. H2O in blue and Africa Foundation in orange convey the idea of water and land. (b) H2O consulting - The name "H2O" serves as the identification of the consulting firm "H2O Consulting" but with no specific relations to water in their activities, etc. (Web services). The wave adds the idea of dynamics, growth and analysis. (c) The fair Wasser Berlin International uses a very simple and modern representation of 2 small rounds interconnected to a bigger one showing the two atoms of hydrogen and the one of oxygen in two dimensions. (d) The logo of Eau de Paris plays with the representation of a human form, three drops of water with two of related size and one bigger in the common representation of a water splash. This image is a clear integration of different dimensions of the representation of water (chemical, human and nature).

Box 3.1 H2O TV Show, playing with cultural codes for an international branding, etc. The role of water-based narratives and mythologies

Organisation:	ZDF Enterprises
Creation Date:	2007
Budget:	\$500,000/episode
Target Group:	Teenagers
Audience:	250 millions viewers, 120 countries

Content/Approach

Three sixteen years old girls are changed into mermaids after an adventure in a mysterious island. A magic spell transformed them into mermaids with powers each time they touched water.

A story of Branding

The show is distributed by Disney Channel being very popular among teenagers. The relations with water introduce the idea of protecting the resource. The focus is on mermaid but the Australian Gold Coast as well. However, despite water-based narratives, the show appears to promote leisure activities and touristic places of the area.

The branding "H2o" with the music, references to American and European lifestyle explain the success of the show worldwide. The fantasy e.g. myths of mermaids and the focus on the stories of the girls allowed other cultures to appreciate and identify to the show. As such, references to water, environment and mermaids played an interesting role in the success of the show in the US And Europe (Ward and Potter, 2009).

Box 3.2 The role of "Mascot": Water Heroes in targeting younger generations

Organisation:	State of Colorado
Creation Date:	2012
Budget:	\$463,800
Target Group:	Children, teachers, businesses, facility
Audience:	Target: 260,000 inhabitants In 2012: 17,000 individuals, 500 businesses, 3,000 visitors to websites

Content/Approach

The program "Keep it Clean" implements education programs in school, organizes events, distributes brochures but also, delivers label to business. It targets the protection from storm water in Colorado State.

Jo, H2o

Jo is a water drop. Round with small legs and arms, he wears glasses as a joyful serious student or a teacher. Jo is in light blue as symbol of purity. The name of H2O is synonym of

Box 3.2 Continued.

this purity and the qualities of the molecule of water. Since the program aims at preserving from pollution from storm water, Jo H2O needs to be protected. His shape and his simple design particularly please children. It helps teachers for their presentation. During events, Jo H2O becomes a real mascot (human size costume). The intervention of Jo is a way to convey messages from the program perspective and as if, water, the pure H2O molecule was talking to the target public. The name is complementary to the name of the program "Keep it clean". Jo, H2O is the water drop and similarly, he "has been working hard to protect water quality, to promote and create action about water quality". www.keepitcleanpartnership.org

The use of a mascot is common and many are recognized and known in the water sector such as Hai Bao from the International Exhibition in Shanghai (2010) or Fluvi from the Zaragoza exhibition in 2008 (Figure 3.4). The two characters have been presented as the embodiment of their exhibition being the interpretation of the slogans (Hai Bao for "Better city, Better Life") and with their personalities (Fluvi is "positive character, purifying, cleaning and nourishing" and Hai Bao is "embracing friends from all over the world"). Being characters with their own personalities and their specific qualities, mascots are branded heroes accompanying the program.

Figure 3.4 From left to right: Fluvi, the mascot of Zaragoza Exhibition 2008 & Hai Bao. The mascot was adapted to other cultures being dressed with various national costumes e.g. Spanish, Hawaiian or American costumes (*Source*: Wikipedia).

For the Beijing Olympics 2008, BeiBei was one of the five mascots of the Olympics Games. There are also several national and local initiatives with their own mascot and they usually target children. They accompany the awareness campaigns for events and to identify the programs. Few examples: BeiBei, Beijing Olympics 2008, China; Water Wally, Singapore Public Utilities Board, Singapore; Kanaat El Mae, ONEP, Morocco; Pronasar, Peru; and Grifo, Aguas del Norte Salta, Argentina.

The European Commission also used a small hero to accompany its campaign targeting children and households, "Generation Awake". The main hero is named Water Maniac Walter and is the representation of bucket filled with water. To introduce the hero and website, a TV spot depicts a psychologist listening to someone obsessed by saving water. The viewer listened to the complaints that seem "normal" as the doctor states. When the doctor spills water accidently, Walter jumps to save water frenetically. The message plays a double meaning to encourage people to be more careful, more "maniac" but not as maniac as Walter who appear to be too "neurotic". This characteristic of the hero makes him funnier and facilitate the acceptation of the types of new behaviours the website is promoting. In its press released dated May 22nd 2012, the EU announced that the video with Water Maniac Walter has been viewed by more than 1 million viewers and the campaign website generated over 200,000 page visits.

3.3.2 Water Colours

Water has no odour, no colour and no taste. As such, water could be represented in any colour, etc. However, water is usually represented in blue according to Pastoureau (2006). For the seawater, iconic representations will usually use blue in particular for maps. The first historical maps by Ptolemee, considered as one the most famous geographers of Ancient time, used blue for oceans, rivers and lakes as shown by maps reconstituted during the 15th century.

Mosaics from the 6th century or maps from explorers (for instance, Portolani) usually use blue for rivers (Swift, 2008). Rivers and lakes were also traditionally painted in blue in the first maps and blue remained the colour to identify water and water resources. Pastoureau (2006) underlined that blue expressed "cold" colour when symbolizing water, in particular for mineral or tap water. Blue is opposed to hot, warm representations and colour such as orange.

In terms of symbolic, the heritage of the blue colour showed four stages of significations for blue (Pastoureau, 2006). In his book, Pastoureau states that first, blue remains absent or only mentioned with discontinuous characteristics and imprecise definitions. From Antiquity until the 12th Century, blue was not included in the system of colour where white, red and black were the main important colours. During medieval times, blue became the colour to paint the veil of Virgin Marie in religious icons; it conveyed the value of her virtue and virginity. It also became the symbol of royalty in France starting in 1130 with the Capet Family. Blue was quickly used in many royal and noble coats of arms. These two religious and political affiliations probably facilitated the use of blue as a symbol of morality. Blue became the symbol of honesty. During this period, many colours were listed as forbidden or authorized for special use in particular due to the Protestant Reform. Since blue is not a major colour, rather "invisible", it stayed freely used and continued its ascension. Starting in the 18th century, blue was known as the favourite colour of European populations. Blue, when integrated into the primary colours (red, yellow and blue), conveyed new significations and was widely used in flags and political movements. Today, accrording to Pastoureau (2006) blue is a consensual colour synonym of God, tranquillity, reverie or romance along with justice, peace and many official international organisations (the UN, UNESCO, the European Commission), etc. Due to its popularity, Pastoureau even pleads that blue has became invisible again as much as white (Pastoureau, 2006).

Regarding other colours, white and green are the two other relatively frequently used colours to depict water. White has long been used as a colour to symbolize God as the "light of God". In religion, as in antique times, white is often the primary colour. Aristotle considers white as the first colour in his classification, the church used white for clothes; to embody the Holy Spirit; and later to signify purity and virginity. White was also the colour of innocence. White later became the "non-colour". The demonstration by Newton that white and black are not colours definitively sealed the significations of white. According to Patoureau (1992) white is a colour and the only one that kept the same significations over time: innocence, divinity and purity.

Green, in reverse, has a more tumultuous history. Green has long been disregarded since it was the result of mixing colour and very instable colour. As such, green became the representation of destiny, fate or luck. It was used for playing cards and money. Early Islam most likely introduced the representation of nature in green but until the 18th century, nature was painted through the four elements than in green. Today, green is the symbol of the planet and honesty, where the word "green" is used to qualify an ethical initiative or sustainability. The expression of "green washing" will actually confer its historical ambivalence being synonym of luck and bad luck, gaming and instability. (Pastoureau, 1992)

Looking at the "colour of water", water can be associated with different colours in order to specify a type of water:

1 White water: frothy water

2 Blue Water: (deep) the open sea (Blue water) surface and groundwater
3 Green Water: rainwater stored in the soil or vegetation
4 Red Water: any of several cattle diseases/Red tides: harmful algae bloom
5 Greywater: household wastewater from bath, shower, washing, cleaning, etc. slightly contaminated
6 Brown Water: colour water from pollution in particular in coastal zones
7 Black Water: Waste, water containing faecal matter and urine higly contaminated by bacteria

The different colours except for red water; are referring to different states of water in the water cycle. They respect the classification of colour from lighter colour to darker water along with the water and wastewater cycle qualifying the visible pollution of the water liquid. We combined the spectrum of colour with our mapping of words qualifying water. This interpretation clearly shows how colours are related to significations held both in words and choice of colour when setting communication on water (Figure 3.5).

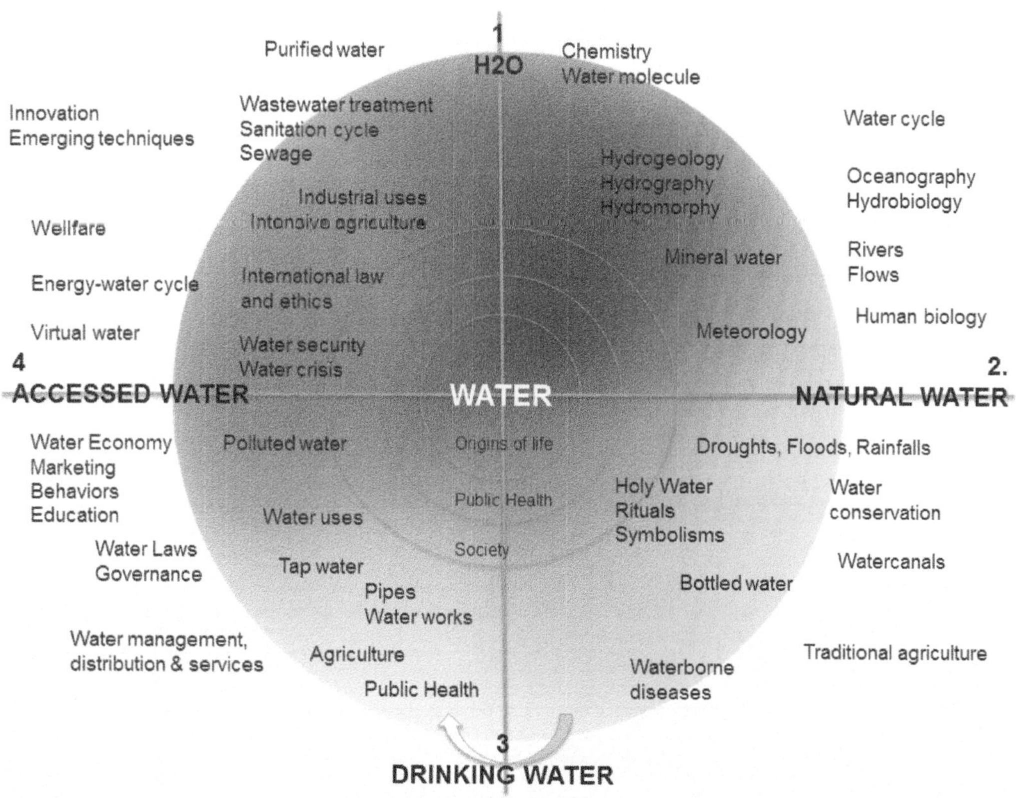

Figure 3.5 Combination of words with the spectrum of colours. As a reminder, the "visible spectrum" contains blue, green, yellow, orange, red, violet and indigo. Our central colour is indigo-blue-green (upper and right half) opposed to yellow-orange-red (lower and left half). From H2O to natural water, colours range from different type of blue. From natural water to drinking water, green is prominent. Yellow, orange and red represent drinking water to accessed water. After accessed water back to H2O and including latest water technologies, colour are usually pink, violet, indigo or back to dark blue.

According to our figure, the combination of terms with colours shows the signification given to colour when representing water.

- The Blue tends to refer to chemical and natural water. Water is considered in its pure state with mysterious attributes. The combination with green shows the relationship with the environment. The combination with purple and pink embodies its chemical values (Ultra violet).
- The Green refers to the environment, the Earth.
- Yellow and Red will refer to energy but also to civilisation, access, treatment. In most cases, it will confer the idea of soil, sludge, and treatment.

  The combination of colour with word actually highlights that symbolizing water in blue is referring to its pure state where as red, yellow or brown will eventually suggest treatment. The green acts as link between purity, nature and human involvement. In general, purple and pink are used by organisations focusing on human relationship i.e. initiatives or programs that are not offering the services of water treatment. The use of blue with small use of other colours (red, pink, yellow) is also dominant in networks or programs. Most utilities, ministries, NGO or humanitarian organisations are using blue. Mineral waters are breaking the codes using a large variety of colours but the majority combines blue with another colours.

In the water sector, the use of blue in logos and campaigns is clearly dominant. The study of 500 logos from 80 countries shows an overwhelming use of blue. On the total of 500 logos, only 74 logos do not use blue or green, this represents 15.3%. (See Box 3.3: Statistical details on the study of Logos, Figures 3.6 and 3.7).

Box 3.3 Statistical details on the study of Logos.

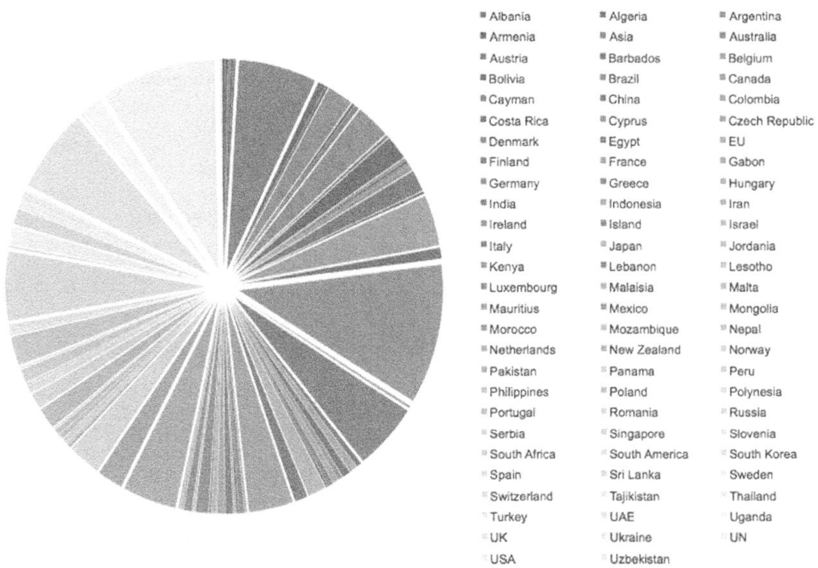

Figure 3.6 Geographical distribution of Logos. The countries most represented are France (10%), USA (10%), UK (6%), Australia (6%), Germany (5%), Spain (5%), Japan (4%), EU

Box 3.3 Continued.

Projects (4%), the Netherlands (4%), Canada (3%), Norway (3%), Belgium (2%), Denmark (2%), Portugal (2%), New Zealand (2%), South Africa (2%), United Nations (2%), China (2%), Austria (1%), Finland (1%), Hungary (1%), India (1%), Italy (1%), Israel (1%), Luxemburg (1%), Morocco (1%), Romania (1%), Russia (1%), Singapore (1%), Sweden (1%), Switzerland (1%). Others countries represent less than 1% of the study.

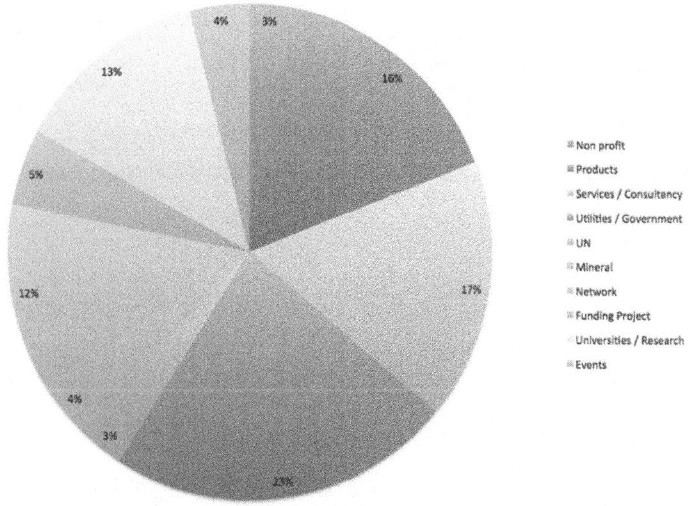

Figure 3.7 Activities distribution of organisations/logos. Most of organisations are utilities or governments (23%). Organisations providing water services or consultancy services represent 17% and organisations selling products/devices represent 16%. Universities and Research centres count for 13% and Networks for 12%. Other organizations include non-profit associations (3%); UN organizations (3%); Mineral bottled waters (4%); Research funded projects (5%) and Events (4%).

Many organisations explain their choice of blue or green for their symbolic significance. For instance, the ANA (Autoridad Nacional del Agua) in Peru, uses green and blue as references for farmers. The green represents the Earth and the blue, water. The shape is comparable to the "A" of ANA but also to a hill. Peru is a hilly country, the green serves the organisation to convey a spirit of security and the preservation of the nature. The blue conveys the special relations of people with water, its calm and also refers to the natural water resources of Peru[2].

Among those 74 logos from the water sector, 62 logos are from environmental companies or universities with activities in water resource management. Only 12 logos are original and are using other colours (2.5%). Two logos are actually from the world's two biggest private companies providing water services e.g. SUEZ Environnement and Veolia Water.

The combination of words with colours to symbolize activities shows the influence of symbolism and significations on water organisations. Combining water word mapping with the use of colour highlight the

[2]See www.ana.gob.pe, Manual de Uso de Marca

importance of symbolic representations and influence. This representation also confirms the similarities in representing water. Water organizations are following the general consensus of the white and blue, or blue and green association to symbolize their activities in visual representations such as logos as seen in Figure 3.8.

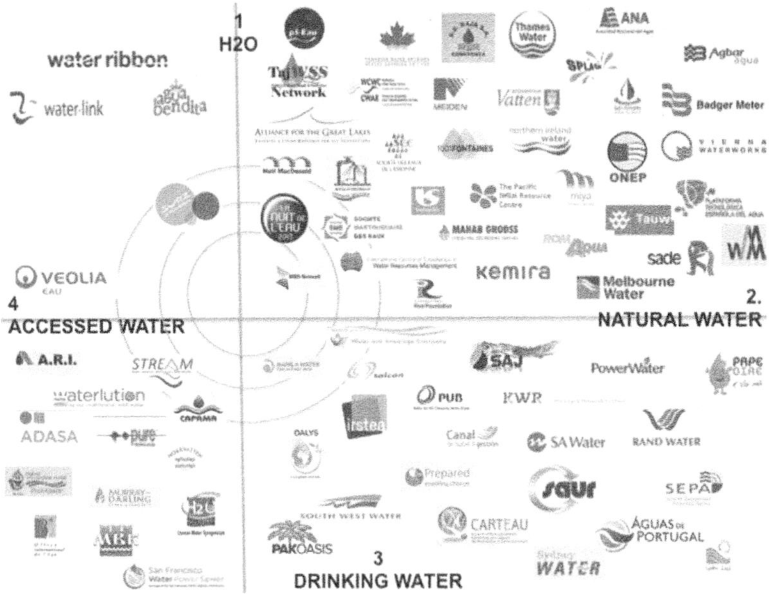

(a) Logos of water organisations, a domination of blue and green colours.

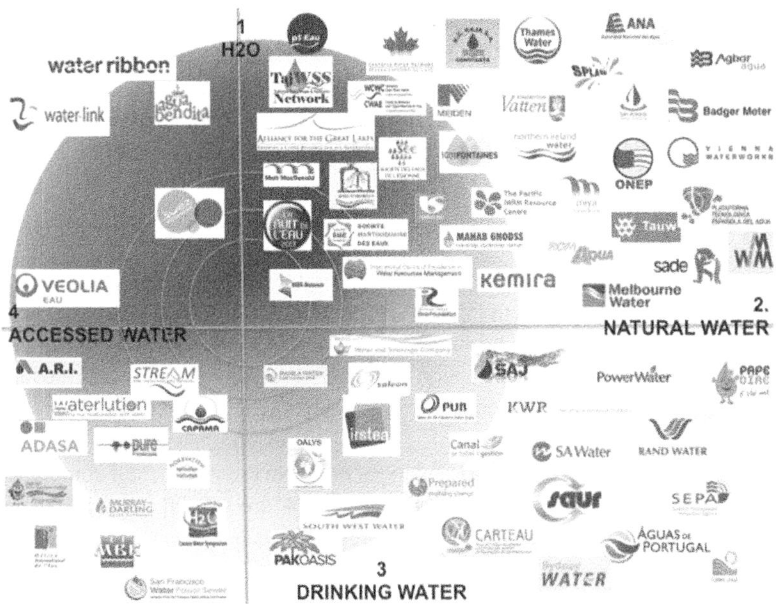

(b) Combination of logos and the spectrum of colours.

Figure 3.8 Combination of logos along the water cycle of significations

The domination of blue (from light to dark blue) refers to the symbolism of water as natural resource, its purity.

3.3.3 Water Design

Apart from colour, other major trends can be found in the historical representations of water in ancient cultures and arts, the design of logos, in pictures representing water and design of brochures or websites.

3.3.3.1 The Drop

The word "Drop" has many meanings in particular, "to drop". It includes to fall, to arrive unexpectedly, to come to an end, to give birth, etc. Those significations reassemble the many symbols related to water. In religious beliefs, the drop is the symbol of fecundity and is related to the dew, holding strong symbolism in the Jewish religion (for agriculture) or in the Bible as the rain that falls on agricultural lands but as the representations of God's spirit in the Old Testament. In Indian tradition, the drop brings humidity and hydrates living beings. The drop is intrinsically related to nourish the Earth and help humans in their cultivations. The drop symbolizes individuals and the community, fecundity and death, rain and resource. The Drop is usually drawn with the same shape of an egg. It can be represented in a "splash" or a drop falling in some water (Figure 3.9).

Figure 3.9 The drop, universal individuality and community. For example, the logo of the European Water Community.

The repetition of the use of a drop by water organisations shows the importance of symbolising water resources. The drop particularly refers to the uniqueness, the vulnerability and the importance of water. The drop in many references is the drop that changes courses of actions or the last drop. The name and the design allow some creativity but convey the same significations and symbols. Several organisations built their campaigns playing with symbols, names and references (Box 3.4).

Box 3.4 The Drop, a concept of one and several or how linking community and individuality to water.

In 2011-2012, the UNEP and the UN Regional Information Centre (UNRIC) organized a major online contest to invite artists, designers, professionals or any persons engaged for water to develop a poster add representing the concept of "Drop by Drop" (www.dropbydrop.eu). They received more than 3,500 ads from 45 countries. The slogan was "The future we want" and the print advertisement should inspire "others to preserve water, now and for future generations". The competition also wanted to inspire the European public in a "positive way".

Looking at the two winners, the choice of the public represents the embryo in a water drop focusing on the notion of life. Over the 3,500 ads, 1,156 are classified under "life" e.g. 33% of participants submitted an ad using the slogan "life". The choice of the public reveals how people perceive water, as a drop of life and the symbol of maternity (Figure 3.10, see also Chapter 6, Figure 6.4).

Box 3.4 Continued.

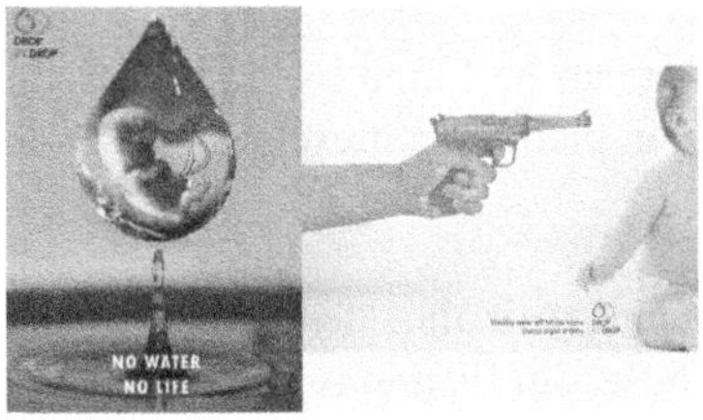

Figure 3.10 Winner of the Public Award, "No water, no life", Toprak Onur. He gained 11,000 over 80,000 votes (*Source*: UNEP, 2012).

The winner of the Jury's price however didn't refer to a drop but to life (Figure 3.11). It represents a hand pointing a blue plastic toy gun to a baby.

Figure 3.11 Winner of competition: Daniele Gaspari, Wasting water will kill the future (*Source*: UNEP, 2012).

Several criticized the choice of the Jury. They saw a strong relation with a former UNICEF ad dating from 2008.

The confusion by the public is particularly revealing. There is a number of divergent messages on the water cause that are contradictory. If the image is actually similar: a gun pointing at an innocent human being, the message is different. For both images, the gun is a toy showing the amusement, the fact that humans are playing with water rather than protecting it. In the first, the gun is blue which brings to the representation of water. The first picture implicates the future (baby) and wants people to protect water resources whereas the second picture advocates for clean water access.

Later in May 2012, the European Commission released a TV spot titled "If water could speak". The spot shows sand timer filled with water, which visualize the idea of the "drop by drop". A feminine voice speaks as if she was water. She remembers that water is a treasure: "they named me the diamonds of the God" which is a religious reference. Some chemical aspects are also underlined (transparency, purity, freshness). The narration continues and accelerates thanks to the music after a key phrase: "I am Water, the source of life". It's a way to introduce the different water uses, to introduce key messages on waste and the need for its preservation while some pictures of women or men are shown inside the timer. The key message is "every drop of mine counts".

The spot ends with the visualization of the last drop and the writing of the slogan accompanied by the key message: "We need to rethink the way we use our water".

Box 3.4 Continued.

Exanimating the blogs and discussion after the contest, several people complained about the lack of originality from the jury in choosing the winning photography and several also believed that the TV spot used their ideas ("drop of mine", "sand timer'). It showed the difficulty to propose innovative message on such a consensual topic: the drop as the representation of scarcity, water preservation and future generations.

The "drop" is at the core of many other communication campaigns. The naming of "drop" will play an important role to convince people to become "the drop that changes everything".

The US non-profit One Drop developed its communication strategy around this idea of one individual bringing change. Their slogan for their TV spot is "It starts with just one drop". It gives the voices to two young girls (a teenager and a kid) telling how they became aware of the water crisis and initiated changes in their circle that eventually, brought global change.

> "I tell him and he tells the school. Teachers tell the classes, students tell the families. Families turn to cities. Cities become countries. Countries make global changes."
> (*Source*: One Drop, 2013)

The last campaign is based on the same concept but addressing the Poker Community with a campaign called "All in for One Drop". The winner of the World Series of Poker shares his experience with One Drop. The campaigned is designed playing with the codes of "poker" including the drop framed in ace of spades. A dedicated website presents the partnership and the approach to the community also showing the importance of one to build a community.

Other non-profit uses the term "drop", for instance in other countries. In France, the non-profit association Solidarités International named its online petition "your drop of water" (votre goutte d'eau) or a non profit association association in Haiti is named "Water Drop" (Gota de agua. They all play with the idea one drop, one person making the differences for others and for access to water resources.

3.3.3.2 The Wave, The Flow and The Vessel

The wave refers to the sea and the marine imaginary. Many have studied the different characteristics of the sea in religious and symbolic contexts (Bachelard, 1942; Gritti, 2001). Waves have the double meaning of being calm and powerful, passive and active. The ambivalence of the wave is related to the ambivalence of the sea. It applies to river flows that are both agreeable and dangerous. They are embodiment of dynamics, force and nature itself. A related representation would be the vessel introducing the capacities of men to conquer those elements, to sail and to advance into waterways or beyond oceans. Mermaids can be found in this context to accompany materials and campaigns.

It is less frequent to find communication campaigns based on the wave or the vessel to only address drinking water. It however can bring awareness on coastal zones pollution and be a recurrent figure in the communication of cities by the sea representing today, 60% of the world population. Numerous logos use the design of a wave or a flow to relate to both the sea and the river (Figure 3.12). They are usually linear waves in different blue or/and green.

Figure 3.12 Waves, the ideas of flow, movements and dynamics. For example, Agbar agua (Spain).

Similarly, vessels are reminiscent of the coast or the bays. They are usually found in logos from utilities on the sea or rivers' basins with strong harbour activities (Figure 3. 13).

Figure 3.13 The Vessel, the iconography for coastal zones, harbours and the historical patrimony. Example: The logo from Eau Seine Normandie, the Agency for Normandy and Seine Basin which covers a coastal territory including several harbours such as Le Havre, a major French harbour.

In communication campaigns, rivers are very popular representations of nature. Calm, quiet, preserved, they intend to show the importance of natural resources. Rivers in water campaigns usually speak for themselves and accompanied by catchy slogans initiating to the preservation of water resources. Sea images or sea animals are also recurrent visuals that will lobby for the same types of discourses for the environment and sustainability. On many occasions, sea adventures are ways to raise awareness on sea protection but also drinking water by providing information on desalinization or coastal protection and the risks of salty water intrusion into groundwater resources. For instance, Lyonnaise des Eaux initiated a partnership with Raphaela le Gouvelo, windsurfer. They developed a pedagogical kit for children including information on seawater, reverse osmosis, desalinization plants, etc. In 2011-2012, the solidarity boat sailed in 10 cities around the Mediterranean Sea to share solutions about water resource management. In 2010, Quicksilver and Roxy launched a campaign using the codes of water as a natural resource (Figure 3. 14).

Figure 3.14 "No Water sucks": how the absence of water creates the message (*Source*: Quicksilver and Roxy). The contradiction strikes between the usual scenery of surfers and the desert. The image highlights the slogan bringing a powerful advocacy.

The first visual remained one of the most successful visuals representing surfers bathing in a deserted wave. It played with the imaginary of surfing and clearly strikes at the eye. The campaigns included other visuals including a boat on rocky hills or empty skiing chair in a desert. A similar

3.3.3.3 The Circle/The Globe

Water represents 70% of the Earth surface, water is naturally represented as an important element of the Blue Planet. As such, the use of the circular form is very frequent in logos as well as the representation of the Globe in reference to geographical maps. There are usually more used by official institutions e.g. non-governmental institutions. Circular shapes are similarly cut by a flow of water cutting the logo into two parts also a conventional representation of the Earth (South/North).

3.3.3.4 The Human Being

If water logos are referring to the Earth due to the importance of water in the Planet, they are also frequently referring the human being. The human body is composed of between 65% and 90% water depending on age. As water plays a key role in human life, the representation of human body or shapes of individuals is also a way to remind the human relationships towards water. It places the individual or the society at the core of the visual representations. In many different ways, the human being can be a water carrier in reference to the historical role of water carriers; historical faces for utilities reminding of their history (below, Stockholm Vatten and seattle.gov). The human can be at the centre of representation to underline the role of individuals (Figure 3.15). Another way is to suggest the human being with modern drawing (water-link) or parts of the body: eyes, hands, or steps.

Figure 3.15 Water as a social good, the figure of individuals.

The representation of the hand or the step is now common and understood as the "print". In the water sector, it refers to the concept of "water footprint" that developed the concept of virtual water.

The other main important and predominant representation related to the individual is water and life. For the second Water Decade, the logo of the UN symbolizes water received by hands. Water and life are a recurrent association placing a being at the centre of the communication strategy. It also allows communicating on key values such as the future, the community (for "all") and for each human being.

3.3.3.5 The Cycle

The concept of cycle is very important in water management resources. The water cycle is at the basis of the work of water professionals. The many iconographies of the natural water and domestic cycle play an important role to bring information and knowledge into pedagogical formats and design. In logos, many companies or events refer to the concept of cycles as the idea of recycling, completion or infinity. It conveys environmental values to protect limited resources but also to turning them into infinite quantity. The PUB, the Singapore national water agency uses this symbol since they are able to close the water cycle thanks to their reuse water technologies (Newater).

3.3.4 Water Iconography

The recent development of visual data on the Internet contributed to an explosion of visualizing water challenges into simple to complex online iconography. The new possibilities offered by the tool allowed many non-profit organizations to illustrate their activities into dynamic iconography, institutions to provide key information about water challenges or many citizen initiatives to develop independent ways to visualize water resources management into key statistics and informational tools.

Those new tools help to explain the water cycle but also to accompany the TV spot in order to describe the program's activities. In many shows from non-profit organisations, interactive drawing clarifies what is said. It is very common to find iconographies to explicate data, activities and why people should support the programs. They represent people, the objective of the actions and the need to join together to have an impact on the challenge.

The popularity of turning data and discourses into visualization and iconographies has brought many organisations whatever their status, to use such data to explain their activities or their cause e.g. militant organization or political initiatives about water.

3.4 WATER SLOGANS

Throughout the history of Water Communication and campaigns, the analysis of discourses, water words and logos, several recurring slogans can be identified. They construct the codes and key messages around water resource management as well as perceptions of water challenges today.

3.4.1 Water is Life: A Common Good

The far and foremost most famous slogan today is "water is life" or "water, source of life". Since the UN launched the Decade for "Water, a source of life", the naming of campaigns is built around this simple and universal bound: water brings life.

The naming is very important and varies from "Water is life" & "Water for life" (two most used), "Water of Life" (mostly Religious), "Water and Life", "Save Water, Save Life" or "Waterlife" (name of Canadian initiative about the Great Lakes (National Film Board of Canada and Primiative Entertainment inc., 2010). There is a wide number of organisations using the slogan water / life including NGOs (UNICEF with ACQUA for life), political and governmental institutions (the EU used this slogan to inform about the European Water Directive); non-profit associations will usually refer to the slogan or the UN Decade to justify their actions and the need for donations; private companies or bottled water including Evian whom launched a program named "Water is Life" as part of its Health and Water initiative.

Many programs will choose a baby or young child to accompany their program about water and life. There is overwhelmingly frequent visual repetition of a child or young adult drinking water from a fountain or a tap with water flowing in his/her hands.

One expression of water and life is the recurring representation of the first stage of life as a foetus. For the UNEP contest "Drop by Drop", 95 visuals submitted associated water with a foetus or pregnancy including the choice by the public (first winner see Figure 3.10 and Figure 3.16).

Figure 3.16 Pregnancy, a recurring figure for water for life in public's perceptions. For the UNEP's contest "Drop by Drop", 95 visuals associated water with a foetus or pregnancy including the choice by the public (first winner) (*Source*: UNEP 2012).

Mineral water brands also play an important role in promoting the links between motherhood, babies and water as a source of life. The most famous brand that built its slogan, identity and campaign around this linkage is Evian. Since the brand introduced babies in her ads, the brand has first developed the concept around the mother, the baby, the life and now, youth. The brand slowly shifts from the mother to youth but the key value remains life in various representations. The different ads by Evian are referring to perceptions on water and health. They are inspired by codes from thermal cures' heritage. They first framed mothers giving mineral water to their babies, and then emphasized the links with the life/pregnancy to focus on babies. Latest ads underlined the value of youth as a key social shared value by the society today.

In 2013, for the World Water Day, Serge Belo, a Canadian artist decided to raise awareness on the need to preserve water, the shortage of water supplies and to the consumption of goblets by constituting a large mosaic made of cups filled of dyed rainwater. He gathered 66,000 cups and draws a mosaic representing a foetus to embody the value of life.

The terms life can also be associated to other words that will refer or remind water such as "wells of life" a US non-profit association; or with water in another language such as "Acqua for life" since the program is supporter by Giorgio Armani, the famous Italian fashion designer. Water and life is widely used by many different types of programs: water access, health, and pollution, etc. It embodies the universality and the human cause. In terms of discourse and messages, water and life refer to water as a common good, a universal bond, and a collective natural resource that should be protected by mankind.

3.4.2 Water for all: Water Access; Sanitation for all: Decent Life

The slogan "Water for all" is also predominant in water campaigns. Many professionals from the water sector actually stated how fed up they were by the use of the slogan particularly during the 90s and 2000s. The slogan became a mainstream with the UN Decade 1981-1991. Today, many expressed how ambitious and unrealistic this slogan was. Many also believed "water, source of lie" as better and more universal choice.

In terms of campaigns, "water for all" is very frequent. The naming "water for all" can be found in many campaigns. Mostly, the use of "water for all" is associated to the general cause of promoting water access to drinking water. Many non-profit associations will build their discourses to promote water access for all but will choose other types of slogans to be different from institutional discourses that are investing in the slogan "water for all": funding agencies, NGOs, private companies usually choose water for all as a systematic reference: World Bank, Asian Development Bank, UN-Water, WSP, UNICEF in partnership with many private companies such as UNILEVER, or Suez Environnement. The slogan "Sanitation for all" became itself a slogan particularly in 2008 with the international year for sanitation for all organised by the UN.

While "water for all" focused on access e.g. supplies; infrastructures; wells; taps, etc. Sanitation for all is more complex. It means access to toilets, to sewage but mostly, to dignity. For women, it means security and respect. The WASH campaigns are particularly strong and very well known among water professionals (see survey on "Water Communication"). Many answered that the WASH campaigns are the most successful campaigns towards the public in the last decade ("Hurry up!", "The Big Issue" and "Sanitation for All"). The strike by Matt Damon or the advocacy by WaterAid also put much visibility on the Toilet Day and the slogan "water and sanitation for all".

In 2013, the Peruvian University UTEC (Universidad de Ingeniería & Tecnología) with the city of Lima built a panel that transforms air into water. With the support of a communication agency (Mayo Publicidad), they branded the panel into a communication campaign to show how technologies can provide solutions to water access. The panel is built in the suburbs of Lima at less than 90 km from the centre where many communities don't have access to water. The panel can provide 96 litres of water per day. The water is free and the panel is accessible to the public and local communities. The campaign to promote the initiative was built on social media and press releases. One of the headlines was "advertisement can produce water". The initiative and communication campaigns won the Cannes Lions International Festival of creativity.

3.4.3 Dirty Water Kills Babies, Children and People: Water Quality

As said Pasteur, we drink 90% of our diseases. Dirty water makes people sick and today kills more people than any other causes of death. This simple message has been widely used as an inspiration concept to many water organisations (see interview with Renaud Doucy). There is no clear slogan as "water is life" or "water for all" regarding the fact linking water and health. There are nonetheless, many campaigns about this issue since it is usually the main challenge most non-profit and NGOs try to address. In terms of recurring characteristics, many can be found: the figure of the baby (Figure 3.17), the colour of the water, the idea of war, and the framing of key numbers.

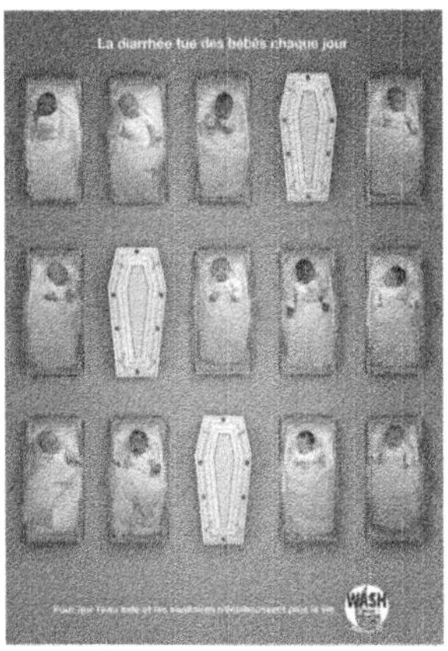

Figure 3.17 Babies, a recurrent figure to visualise the challenge, blame the injustice and claim for actions. Many also use baby bottles in their advertisement such as Helvetas, Charity water, etc. (*Source*: WASH/WSSCC).

Another regular argument is to show how much water can kill compared to famous causes of diseases such as war, AIDS, cancers or nuclear explosion (Figure 3.18).

Figure 3.18 The nuclear explosion, the representation of dirty water killing people (*Source*: Solidarités International; also found in ads by UNICEF or WWF).

Non-profit associations and NGOs are usually orientating their campaign and vision on the message of water and health. Through history, the UNICEF played a key role in promoting the links between water, health, hygiene and vulnerable populations, in particular children. The strong advocacy over the argument "water kills people, water kills children" can engage the public in particular parents or grandparents. They use similar codes and discourses (see 3.2.2. Similar discourses: overview of non-profit campaigns). Today, many non-profits and NGOs tend to develop new codes to get the public's attention.

In 2013, Lifebuoy, an Indian non-profit association that aims at promoting hygiene thanks to the use of soap, launched a commercial that has been widely shared on Internet and is a successful example of building a campaign towards the public. The TV spot showed a man in an Indian small rural city walking on his hands. Slowly, he advances out of the city. Many observers and people joined him by praying behind him. Close to a road, a woman in a bus sees the scene and decides to join the parade. She follows the group without understanding what's happening. The man reaches the temple and meets a religious man. He stops and sits. He offers his respect to the man and tells him his boy has reached 5 years. The woman asks why he made such a fuss about his son's fifth birthday. A man explains that this man has lost 4 children before his son. The woman understands that it's big challenge for children to reach 5 years in rural areas. The commercial ends and reminds that 2 millions of children under the age of 5 die of infections like diarrhoea or pneumonia. Those diseases can be prevented by the use of soap.

The story of the man (his first son is 5 and he goes to the temple to thank God), the solidarity among the village, the vision from outside brought by the woman, etc. The spot builds a universal proximity through the human story of daily life where death and disease are frequent. Many codes are refereeing to Indian practices where the individual act of courage underlines how one can bring change. The discourse is positive and will convince people they should address the challenge of hygiene. The ad has received strong responses with 1.4 millions views in less than three weeks and by more than 7 million viewers on YouTube since February 2013. The Facebook fanpage reached 3.3 millions fans during the summer 2013.

In 2010, the Cannes Lions Festival engaged the public in a contest for WaterAid. The objective was to create an ad that will mobilise policy makers to act to promote access to water and sanitation. One of the finalists created an ad showing glasses filled with dirty water. Each glass was different and related to country's custom with the name of their leaders e.g. a glass of champagne for French President; a glass of whisky for the US President or a cocktail glass for UK prime minister. The ad ends stating, "If the problem was in their glass, they would have already solved it".

This result was comparable to the spot created by Action against hunger for the G8 summit. It engaged high policy makers to fight against hunger by replicating the famous Evian ads with people wearing a T-Shirt representing bodies of babies. The discourses are addressing policy makers and the public. They raise awareness from the public and engaging them to give money. It targets policy makers, institutional organisations or funds to confirm for the need for particular funds. Several non-profit and NGOs function thanks to regional funds, funding agencies, international institutions, foundations, etc. To organize visible campaigns confirms their legitimacy.

3.4.4 Every Drop Counts: Water Ethics

See Box 3.4. The slogan of the unique drop that makes the difference is a way to draw from the individual action to a collective change. The concept of "every drop" means "every drop is one drop for change" pushes individuals to think about their water footprint, their responsibility, and their uses of water. It can mean money to give to bring as advocated by non-profit but it profoundly refers to each of us being responsible to preserve water. It brings a sense of water citizenship, an ethics towards water resources and water cycle management to make it efficient.

3.4.5 Water is Human Right: Water Governance

See Chapter 2, History of Water Communication. The slogan "water is a human right" is rather recent. The lobby to promote the right to water has been very strong. It originated from non-profit organisations and anti-globalisation movement. Today, many public utilities and water agencies will to the value of water as a public good. The movement includes a perception of managing water as a public good rather than a good that can be manage by private organisations. As such, discourses will usually used key words such as 'common good', 'patrimony', "public use", and 'right to water'. In South America, the history with private operators is still prevalent. Many national water agencies are proud to promote their public status in their communication campaigns. They insist on the participatory approach towards the public. For instance, ANA in Peru, has created a specific department to manage disputes related to water and based its work on the Peruvian law stating water is "for public use". The approach of participation will enhance values of community and right to water. It facilitates the engagement by target groups in solving water disputes for the protection of the common good. It conveys the ideal of shared governance and will put the responsibilities on citizens and users to better use water resources.

3.4.6 Save Water: Water Scarcity

Water is scarce; save water for future generations; save water, share showers or drink champagne, etc. There are many slogans related to water scarcity. They are suggesting that water quantity is threatened by climate change, human activities and unequal repartition of water resources. Several non-profit associations also used the argument of water efficiency to mobilise and engage the public in particular, by selling water saver or water timer.

Utilities investigated discourses to encourage customers to reduce their water consumption in particular during the summer. The campaign by Thames Water "We are in drought" remains a symbolic and successful campaign. How come people living in London could ever know water scarcity? London and England are known for regular rain. In 2012, the South-East of the UK had experienced two years of dry weather heading towards water shortage during the summer. Thames Water launched the campaign, they wanted to use simple words "nothing fancy or clever", something that would address consumers in a direct way. They designed the campaign on a simple phrase "We are in drought". The campaign was widely disseminated thanks to outdoor posters, digital escalator panels and underground projections, radio spots, ads in the local press, etc. A survey by Beyond Communications Agency (2012) showed that almost half a percent of Thames Water customers have been using less water after seeing the ads[3]. The company pursued her efforts by setting a dedicated website targeting customers. The website "Waterwisely" is designed as a video game and visitors can meet characters from the community who are using water wisely (www.thameswater.co.uk/waterwisely).

Many other stakeholders will address individuals to encourage lower water consumptions through different arguments: price, climate change or individual responsibility. The EU project "Renew" developed a public TV spot to link water and energy consumption with billing. The project aimed at showing impacts from using devices helping to control water consumption on water and energy bills.

Helvetas (Switzerland) launched a campaign to show impacts from climate change on water quantity. The print ad represents a glass looking as dry bones and skeleton from desert and arid area.

For instance, Campahnia Atletica (Brazil) encouraged its clients to save water as part of their engagement for Sustainable Development. The targeted customers are sporty individuals who take showers or baths after sport. The campaigns illustrated funny situations where a shorter bath or shower could save the person.

3.4.7 Water, The Blue Gold: Water Economy

The discourses over "water wars" are usually brought by the media. Documentaries, headlines, web documentaries or TV shows about water resources will usually focus on wars over water resources. The naming of "wars", "blue gold", "petrol of tomorrow" constructs the idea of water being a source of conflict despite many experts' opinions contradicting that water is a source of war.

Many initiatives by journalists won prices including a nomination at the Oscars (Darwin's Nightmare in 2004). They tend to summarize the issues and bring fears over water resources. A web documentary "Disputed Waters" initiated by journalists from different countries intended to bring awareness on the links between conflicts and climate change. They preferred to name "dispute" rather than "war" to underline the risks over shared basin's rivers. They developed an interactive tool (multimedia project) on 5 major rivers including Colorado (USA); Mekong (6 countries); Nile (11 countries); Jordan (5 countries); and Rhine / Danube (11 countries). They regularly update information, feeds and news in order to bring more in depth information rather than short stories on the management of water resources from the rivers' basins. This initiative shows the importance of the media in providing information to raise awareness of the public but also questions the systematic coverage about "wars over water". The complexity of the water cycle and water resources management supposes training and education of journalists to provide other types of information to the public (See Chapter 7).

[3]The cost of the campaign: £120,000. For the full case study, www.effectivedesign.org.uk

4 Research Areas

As stated, the importance of communication to promote water awareness has been recently acknowledged and recognised. Today, little research is dedicated to campaigns on water or to understand public perceptions. This book aims at identifying key elements to further promote the specificity of communication strategies and campaigns about water cycle management. In this frame, there is a need to develop research on "Water Communication". Starting from a limited survey of water professionals, five research areas for Water Communication need to be further examined and funded throughout the world. Such new research programs should support a stronger integration of social aspects and public perceptions into water cycle management.

4.1 SURVEY ON WATER COMMUNICATION

Based on a survey of 70 professionals from the water sector, *Water Communication* involves engaging its audiences, providing important and updated information on water resources, offering messages and strategies to accompany decisions making on water uses (See Box 4.1).

Since water represents such important challenges, to ensure adequate knowledge brokerage and to influence people's attitudes appear to be one of the characteristics of Water Communication. The particularity of water involves responsibilities and ethics, the respect of common resources but not for the accomplishment of an ideal. Water Communication is characterized by its relationships to daily uses of water, invisible technologies and local knowledge. Many respondents to the survey argued that communication on water is about dealing with local challenges, individual concerns, and human-based information. Water Communication is different from the ideals or discursive utopia brought by information and communication technologies. Water Communication recognizes the universal bounds brought by water as a human symbol. Water Communication however, suggests facing strong individual conceptions and perceptions. It involves sounded campaigns with clear messages, discourses, slogans, targets and tools. Such a conception of Water Communication confirms its fragmentation; it explains the multiplicity of campaigns.

Box 4.1 Results of survey on "Water Communication" (2013)

Overview of the respondents

The survey was completed between January and July 2013.

- 43% Women and 57% Men
- 67% of respondents acknowledged to have experience in communication projects or campaigns with various occupations (see Figures 4.1 and 4.2)
- 27 countries from 5 continents (see Figure 4.3)

Box 4.1 Continued

The survey was conducted on Internet. It included quantitative questions and qualitative answers. Half of the respondents have been interviewed by phone or emails after their completion of the survey.

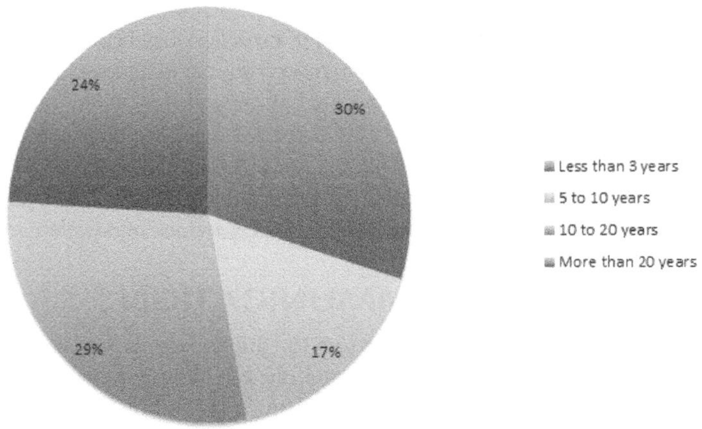

Figure 4.1 Years of experience in the water sector. Respondents have different levels of experience with a relatively balanced repartition: 30% have less than 3 years of experience in the water sector, 17% have 5 to 10 years experience, 29% have 10 to 20 years, and 24% have more than 20 years. 7% of respondents have more than 30 years of experience.

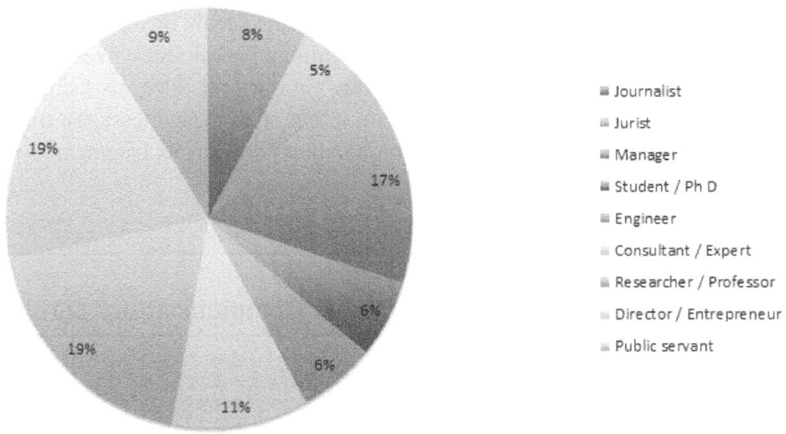

Figure 4.2 Professional occupation of respondents. Respondents represent a variety of occupations from the water sector: 9% of journalists; 5% are law specialists; 19% are managers of water programs or communication/marketing managers; 7% are PhD students; 7% are Engineers; 12% are Consultants; 21% are Researchers or University Professors; 10% are Directors or Entrepreneurs; and 10% are public servants.

Box 4.1 Continued

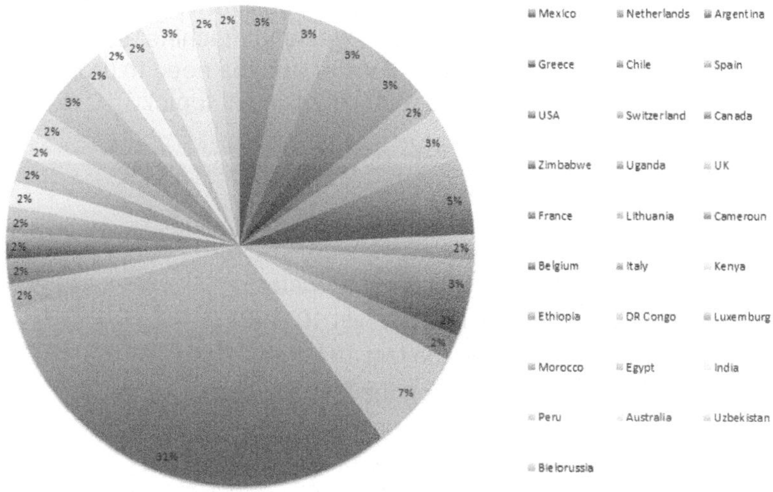

Figure 4.3 Origins from Respondents. The majority of respondents are from France (31%) and Europe (UK: 7%; Spain: 3%; Greece: 3%; the Netherlands: 3%; Italy: 2%), America (19%), Africa (14%), from Asia and Oceania (3.5% for each continent).

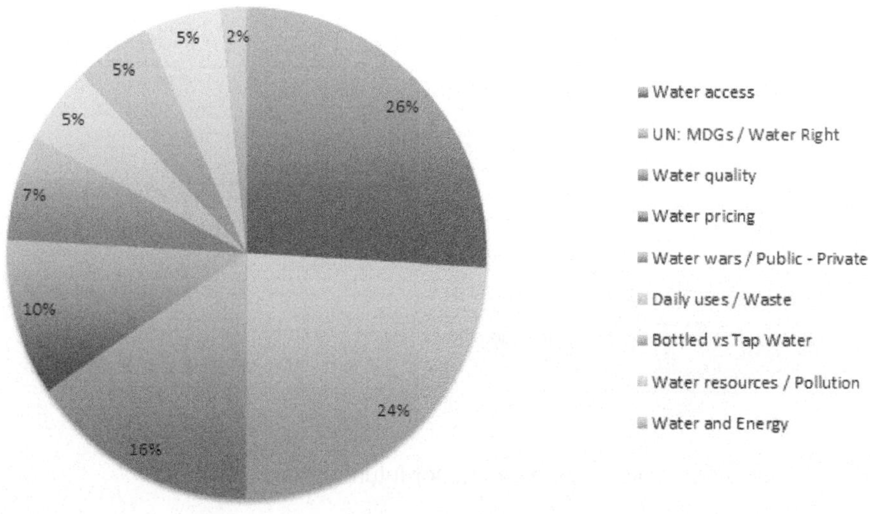

Figure 4.4 The main visible topics about water during the last 10 years.

Respondents consider that water access (developing countries, pumps, lack of access) is the main topic as well as the right of water and MDGs (Figure 4.4). Those discourses are similar since they are addressing the global and local issue of access to water resources and services. Other important topics are water quality (16% mostly about

Box 4.1 Continued

drinking quality); water pricing (10%); water wars and public/private services (7%); daily uses (water waste by individuals: shower, toilets, washing machine, etc.); bottled water (and opposed to tap water, 5%); the pollution of water and the need to protect the Earth, 5%); and water and energy (2% since it's a relatively current topic).

Perceptions of Water Communication

Respondents answered to several questions (Figure 4.5) including:

- Main organisations/individuals communicating on water
- Main issues covered by the media during the last 10 years
- Main visible issues about water during the last 10 years
- The famous campaigns about water during the last 10 years
- The future priority target groups to engage on water
- Their understanding of "Water Communication"
- The main codes, characteristics and references when setting campaigns

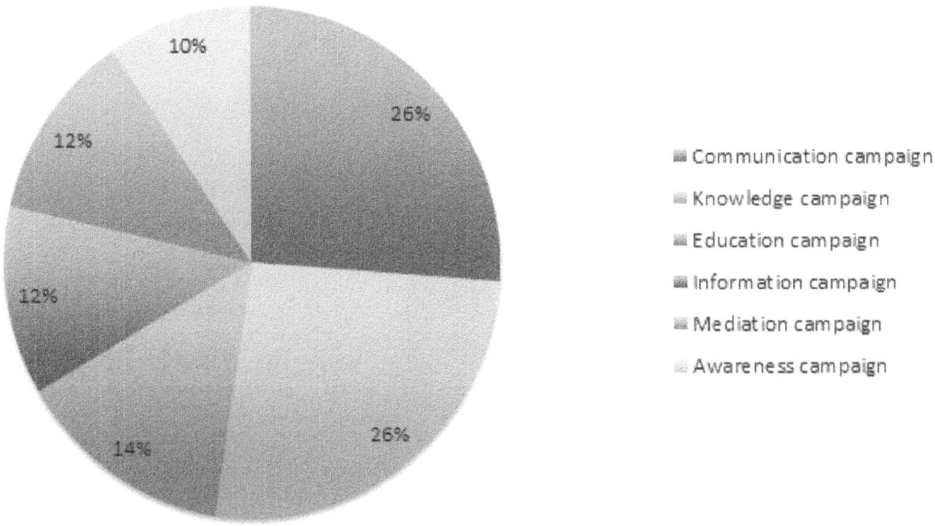

Figure 4.5 Water Communication, needs for future campaigns.

In their responses, professionals suggested different strategies to disseminate information and to engage target groups about water cycle management. Their responses show the importance of values, sharing interactions and to ensure the understanding of the water cycle management and its challenges. Many acknowledged the value of long-term programs through education at different stages of life and adapted to each target groups (Table 4.1).

Box 4.1 Continued

Table 4.1 Target groups for future campaigns

Priority Target Groups for Water Campaigns	1	2	3	4	5
Policy makers	71.20%	26.80%	1.80%	0%	0%
Citizens	55.40%	33.90%	10.70%	0%	0%
Future Generations	51.80%	42.80%	5.40%	0%	0%
Utilities	48.20%	41.10%	10.70%	0%	0%
Media	35.60%	42.90%	17.90%	1.80%	1.80%
Public water services	23.30%	46.40%	19.60%	8.90%	1.80%
Environmental associations	21.40%	48.20%	17.90%	7.10%	5.40%
Humanitarian associations	21.40%	42.90%	17.90%	10.70%	7.10%
Water professionals	21.40%	35.70%	26.80%	12.50%	3.60%
Private water companies	14.30%	44.70%	26.80%	7.10%	7.10%
Cooperation actors	14.30%	42.90%	32.10%	10.70%	0%
Research / Universities	10.70%	41.10%	33.90%	14.30%	0%

The most important group of stakeholders to target during the future campaigns about water is "Policy makers". Other interesting results highlight the importance of campaigns towards Utilities, Media and Public water services.

4.2 OVERVIEW OF RESEARCH AREAS

From the literature review, Water Communication includes the five following categories:

1. Social representations of water
2. Water Discourses
3. Knowledge Brokerage
4. Water Strategies and Campaigns
5. Water Journalism

Each subfield addresses specific needs to communicate on water challenges. They are transversal and overlapping on each other.

4.2.1 Social Representations of Water

This field includes public perceptions, behaviours, consumptions and water marketing. There is a long tradition of studying water perceptions, representations and uses (See Bibliography). The importance to understand public acceptance and to provide information for marketing purposes or for utilities, shows of the relevance of the field. Those studies are however, localized and fragmented. The latest research tends to put a strong focus on public acceptance of reused water; to convince the public to use alternative water resources; to understand consumptions of tap water versus mineral water; to investigate relations between water quality and consumptions behaviours. The figure below by Hurlimann and Dolnicar identified in 2009; identified the research gaps and need to address public water related behaviours (Figure 4.6). The two Australian researchers pointed to the need for research on communication of water.

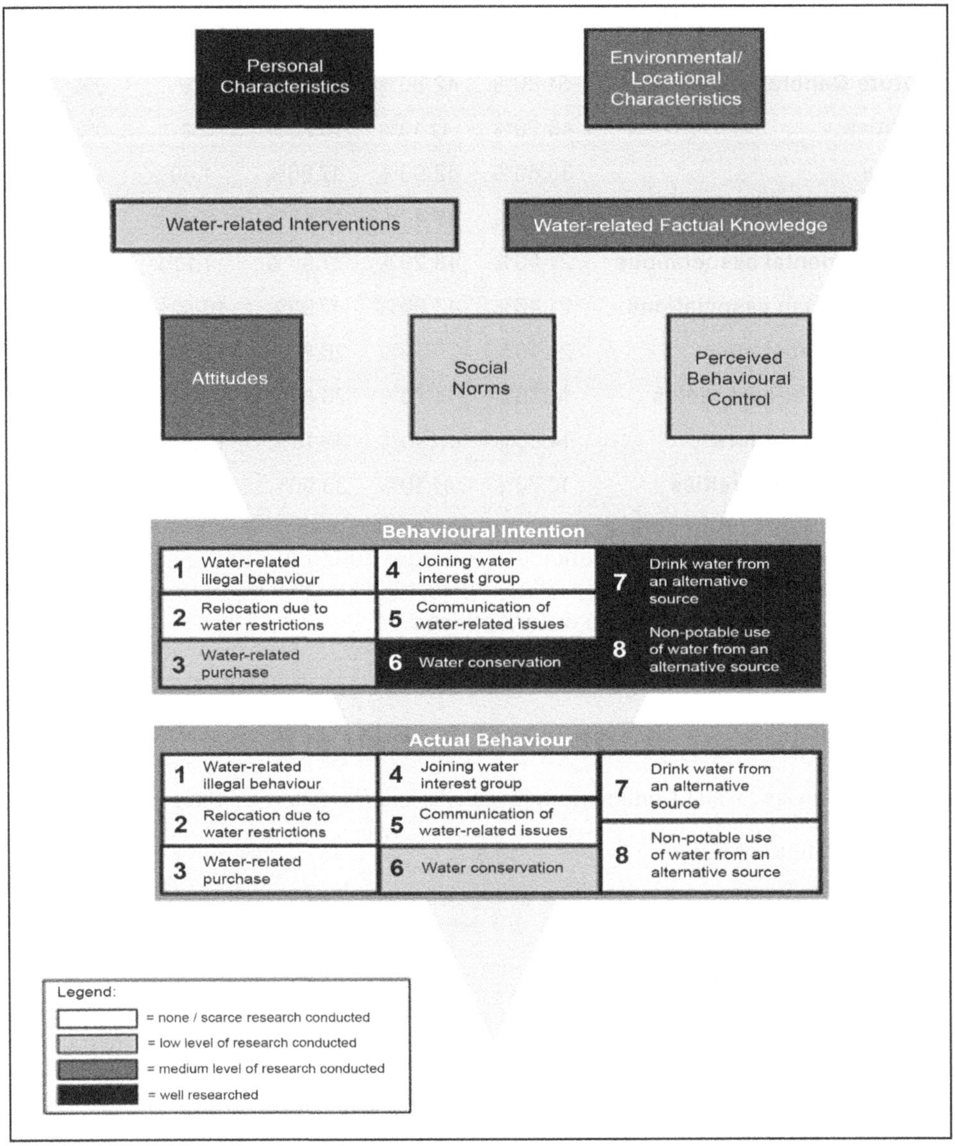

Figure 4.6 Conceptual model and knowledge map for water related behaviours (*Source*: Hurlimann Dolnicar and Peter, 2009: 24).

4.2.2 Water Discourses

This field includes the study of words, linguistics, and rhetoric.

A number of authors introduced the study of discourses as new paths toward understanding water management. The international organizations produced many speeches, statements and texts. They constitute numerous references and "water discourses". Water discourses are defined as discourses setting principles on water cycle management. Those principles are largely repeated and disseminated. They are used as references during conferences, in scientific publications or communication tools and materials. They are the results of consultation processes or produced by organisations holding an influence regarding water management.

Today, more and more authors are publishing about those water discourses. The phenomenon is recent and originated from the development of discourse analysis in international development studies, politics and the analysis of environmental discourses. (See Bibliography) The history of the water sector in relation with the international agenda and the roles played by the UN organisations is further examined in Chapter 2.

4.2.3 Water Journalism

This field covers the analysis of media coverage on water at local, national and global levels. It should include the training of journalists, their perceptions and their participation in major events within the water sector. Traditionally, journalists play an important role in raising visibility on global challenges and providing information to the society. In the water sector, journalists brought awareness and public debates on water management. Several criticized the management by private companies or condemned corruption. The relationship between journalists and water professionals has long been distant. The first decade on water failed in establishing better relationships. The organization of major world forums built new relationships between the media and the water sector. It gave more materials and information to journalists. The organization of the second decade on water, the MDGs and major conferences intensified the visibility and numbers of topics covered about water[1].

The impacts of media coverage over water cycle management can affect water governance; public acceptance of water new infrastructures; people's uses of water; people's perceptions of climate change or water events; etc. Since the media plays an important role in interactions and discourses about water in the public sphere, many organisations developed the training and education of journalists. Several programs are conducted e.g. the UNESCO-IHE, the RNTC, the GWP, the IRC, etc. (See Bibliography)

4.2.4 Knowledge Brokerage

This field covers knowledge transfer, information management and mediation, public participation. Much research is conducted on knowledge management, knowledge translation or knowledge brokerage. The interface of science and policy is particularly important to water practitioners. Facing the huge numbers of platforms, research centres or networks, and the numbers of water libraries, there are many organisations trying to coordinate actions. They aim at providing lists of links, overviews of projects and networks. For example, the Asia Pacific Water Forum Knowledge Hubs proposes an interesting over view of research organization in the region (www.apwf-knowledgehubs.net).

Similar initiatives exist at other regional levels, at the European level (the European Water Community or Wise-RTD), in South America (WOP-See), in Africa (African regional water resource information system – UN Water/Africa). The UN Water listed several initiatives highlighting the importance of sharing knowledge (www.ais.unwater.org). Bielak *et al.*, (2008) clarified definitions between knowledge brokerage

[1] See themes covered by media during the World Water Forum, Hoffman & Hoffman, 2003 and during the UN Water Decade, 2012, www.un.org/waterforlifedecade/water_cooperation_2013/

and knowledge translation before analysing the evolution of science communication at the National Water Research Institute. They provide important principles and examples of activities to address the "multi-sector audience of water policy and programme practitioners" (Bielak *et al.*, 2008: 5). Literature about knowledge brokerage in the water sector is well developed and documented (see Bibliography). Some aspects about knowledge brokering and science policy interface are further studied in Chapter 4.

4.2.5 Water Strategies and Campaigns

This field covers the communication from organizations and actors. It involves the analysis of marketing strategies, customer services, public and institutional relations, communication materials, campaigns, internal communication, events, etc.

Two major aspects should be highlighted: internal and external communication. They are building the main aspects of strategies and campaigns that this research area should be addressing. Many examples of successful internal communication campaigns within the water sector serve at conveying essential values for the building of the organisation and for the building of the water sector. As a broad sector including a large variety of actors (See Chapter 5), the water sector needs itself internal communication. Major events or key concepts such as interfacing (see Chapter 8) are constructing tools, methodologies and processes to the setting of internal communication in the water sector. At the external level, many examples of campaigns and the different strategies adopted by water organisations are also provided in this book: non-profit organisations (Chapter 7), the UN Galaxy (Chapter 6) or utilities (Chapter 8).

The main importance of water campaigns is related to education. Many studies analysed the different public in particular, how to address children and younger generations. They surveyed the importance of public information and public involvement. They further consider the importance of public participation, methods and activities. Another important aspect is related to marketing. Many publications are analysing public perceptions to develop new customer services or to answer needs and water uses. Most of those publications are based on national, regional or local surveys. They provide in-depth analysis of people's perceptions of water related challenges. They cover many topics: drinking water quality, water resources pollution, water legislation and infrastructures, water management by the cities, etc. They usually focus on attitudes, behaviours and perceptions at the level of the households. Guides, survey or scientific publications are developed to accompany the use of this material. Most of them intend to promote the "value of water", the need to engage the public, utilities and investments. They provide examples and best practices to encourage more actions from organizations and stakeholders to address water related challenges.

In this subfield, marketing plays an important part as well as survey and analysis of the effectiveness of campaigns. For instance, DEFRA has been leading many collaborative projects as well as publishing many guides and materials. The UK Department for Environment, Food and Rural Affairs (DEFRA) participated to expand campaigns towards the public based on the understanding of their awareness, expectations, needs and leading multi-media campaigns on specific and targeted uses of water. Other national authorities have further developed this approach. Many researchers also contributed to analyse people's behaviours applied to the water cycle management. For instance, Howarth and Butler (2004) provided the analysis of people's attitudes on water conservation showing different steps from awareness, to interest, desire and then action.

5 Overview of Main Senders

The world of water is famously fragmented, diverse and complex. This state is due to the variety of organisations having a role in water management. This overview provides a selected overview of organisations with important communication activities or playing a role in disseminating messages. These key "senders" of water messages are divided into seven categories:

1. International policy actors
2. Inter-water actors
3. Research and companies actors
4. National, Regional, Local and Cities actors
5. Humanitarian actors
6. Water figures
7. Major events

5.1 INTERNATIONAL POLICY PLAYERS

International policy players gather organizations with a global influence on the global political agenda.

5.1.1 The UN Galaxy

The United Nations is the most symbolic of the global order created by the international community after the World War. Resolutions voted by the organization since 1946 are a good reflection of the importance of water in the UN activities (see Chapter 2).

5.1.1.1 UN-Water

Today, the main UN organization dedicated to water is the UN-Water. It includes different activities and one specialized on communication: UN-DPAC. As mentioned, the UN-DPAC holds an interesting role in promoting communication approach. The UN-Water also provides key documents: WWAP (World Water Assessment Program, WWAR. The organisation developed an information centre online. It also developed several communication tools for events and the implementation of the UN Decade on Water.

5.1.1.2 Other UN organisations

Before the creation of the UN-Water and until today, many other UN organisations cover water in their field of activities. They are including water in their programs according to their specific missions. They are all members of UN-Water.

- CBD - United Nations Convention on Biological Diversity
- FAO - Food and Agriculture Organization of the United Nations
- IAEA - International Atomic Energy Agency
- IFAD - International Fund for Agricultural Development

- UNICEF - United Nations Children's Fund
- UNCTAD - United Nations Conference on Trade and Development
- UNCCD - United Nations Convention to Combat Desertification
- UNDESA - United Nations Department of Economic and Social Affairs
- UNDP - United Nations Development Programme
- UNECA - United Nations Economic Commission for Africa
- UNECE - United Nations Economic Commission for Europe
- UNECLAC - United Nations Economic Commission for Latin America and the Caribbean
- UNESCAP - United Nations Economic and Social Commission for Asia and the Pacific
- UNESCWA - United Nations Economic and Social Commission for Western Asia
- UNESCO - United Nations Educational, Scientific and Cultural Organization UNEP - United Nations Environment Programme
- UNFCCC - United Nations Framework Convention on Climate Change
- UN-Habitat - United Nations Human Settlements Programme
- UNHCR - United Nations High Commission for Refugees
- UNIDO - United Nations Industrial Development Organization
- UNISDR - International Decade for Natural Disaster Reduction
- UNU - United Nations University
- World Bank - The World Bank
- WHO - World Health Organization
- WMO - World Meteorological Organization
- UNWTO - World Tourism Organization

The communication of the UN on water is dense and diversified. It includes various websites for events (Water Decade, World Water Day, Water Year); for resources (UN-Water Library, UN-Water website, UN Water Videos); or for specific topics (for example, UN Water emergencies).

The Water Supply and Sanitation Collaborative Council (WSSCC) is a global multi-stakeholder partnership and membership organization that has been formally recognized by the UN in 1990 to complete the work started during the International Drinking Water Supply and Sanitation Decade (1981-1990). They are very active on campaigning about Water, Sanitation and Hygiene (WASH).

The UN Launched the CEO Water Mandate in 2007. It assists companies in the development, implementation and disclosure of water sustainability policies and practices. This initiative is particularly interesting for business actions and strategic planning.

5.1.2 World Water Council

The World Water Council was founded in 1996; it is "an international multi-stakeholder platform". One of the main activities of the WWC is to organize, the World Water Forum every three years in order to reach its mission and vision of promoting awareness, building political commitment and trigger action to better water resources management and responses to global water challenges. Between 1997, the date of the first World Water Forum, and 2013 when the last World Water Forum was organized in Marseille, the number of participants raised from 300 to almost 33,000 participants. Each new forum gathers more and more participants, offers more thematic sessions or side-events, or conveys key national policy makers (Presidents, Ministers, European commissioners, etc.) In 2013, the city of Marseille in collaboration with the WWC and the World Water Forum of Marseille helped to organize the first Alternative World Forum, the FAME. Answering to arguments from many opponents to the WWC on the role of private companies and the World Bank in the WWC, the FAME wants to offer alternative visions and examples of water governance.

5.1.3 Funding agencies

The most active agencies in the water sector are: the World Bank, the African Development Bank (AfDB), the Regional Coordination Committee (RCC) & Rural Water Supply and Sanitation Initiative (RWSSI).

Many initiatives, funds, national organisations funded international programs on water cycle management, here a non exhaustive list of some actors:

- National Funding agencies with water programs: USAID (US), AFD (France), SIDA (Sweden), GIZ (Germany), SDC (Switzerland), WRDF (Water Resources Development Fund, Ethiopia), JICA (Japan), Spanish Cooperation Agency for International Development (AECID), CONAGUA (Mexico), etc.
- Foundations focusing on water: Water Reuse Research Foundation (US based), the Global Water Foundation, Water Education Foundation, etc.
- Foundations with programs on water: Gates Foundation, the Conrad N. Hilton Foundation, Nationale Postcode Loterij (the Netherlands), Arab Fund for Economic and Social Development (Kuwait), Pro Victimis Foundation (Switzerland), King Baudouin Foundation (Belgium), Welfare Association, Oak Foundation, Narada Foundation (China), Fed Funds (EPA, USA), Smart Water Fund (Australia), Foundation Chirac (France), etc.
- Private companies foundations: Fonds Initiative (Suez Environnement), Foundation Veolia, Thames Water Trust Fund, United Utilities Trust Fund, etc.

5.1.4 The European Union

At the European level, the European Commission counts several Directorates with Units working on water resources including Environment, Research and Innovation, Agriculture, Information and Communication, Climate Change, Enterprises, Energy, Cohesion and Regions, Cooperation. Several funding programs are providing funds for water activities: the 7th Framework Programme, Life+, Competitiveness and Innovation Framework Programme, Water Facility, European Regional Development Funds.

The main important EU initiatives on water are the Joint Programme Initiative on Water (JPI Water), the European Innovation Partnership on Water (EIP), other EU originated initiative such as the Water supply and sanitation Technology Platform (WssTP), the European Water Resources Association, etc.

The other major European Institutions include the European Environment Agency (EEA) with a section on water, the European Investment Bank with a program dedicated to water

Other networks and organisations included the EIONET, the European Water Association (EWA), the European Water Partnership (EWP), the European Water Movement, ACQUEAU (Eureka Cluster), EurAqua, the IWRM-Net, etc.

Several EU funded projects are related to communication: STEP-WISE, STREAM, WATERDISS, AWARE4, PSI-CONNECT4, KNOSSOS, WaterRtoM, SPLASH, the European Water Community, WISE (the Water Information System for Europe), WaterCoRe or European entities from international organisations such as the WHO/Europe Water and Sanitation, the Solidarity Water Europe, etc.

The European Parliament also counts a European Intergroup dedicated to water, the EP Water Group. At the Parliament level, EUREAU is the active main organization. Other organisations include the European Federation of Bottled Water, the European Water Treatment Association, the European Federation of Public Service Unions, etc.

5.2 INTER-WATER ACTORS

"Inter-water actors" are defined as organisations helping professionals and organisations to network, exchange information and best practices. They gather by virtual and physical means. They are many initiatives, networks, professional organisations serving the missions of setting more interactions and dialogues among individuals and companies involved or interested by water management and issues. The inter-water actors are aiming at providing information and encourage knowledge sharing. They usually benefit from public funding or accreditations by the UN organisations e.g. a short non-exhaustive list: the CIEAU (Centre for Information on Water, in France), the GWP (Global Water Partnership, funded by the OECD and focusing on governance and financing), the OIEAU (The International Water Organisation, France), the ISW (the International Secretariat for Water or the SIE in France, Canada with international outreach), the IWRA (International Water Resources Association, France); etc.

Other inter-water actors work thanks to membership fees or funding such as the IWA (the International Water Association), the Alliance for Water Stewardship, the Water Footprint Network, the Global Environmental Management Initiative (GEMI), etc. The World Business Council for Sustainable Development and the International Union for Conservation of Nature issued in 2010 a guide dedicated to initiatives for the water business. It lists a number of initiatives interconnecting businesses, research centres, etc. (International Union for Conservation of Nature, 2010).

Consultancies are excluded but they also play a role in setting interactions and dialogues between organisations and individuals. Network aiming at encouraging cooperation at Basin level such as the INBO, the Euro Mediterranean initiative, Nile Basin Discourse, etc.

It is interesting to note the existence of numerous similar initiatives dedicated to younger generation such as: IWA-Young Water Professionals, Water Parliament for Youth (PMJE), Réseau Projection, World Youth Water Alliance, Rework, AWA for young (Australian network), etc.

5.3 RESEARCH AND COMPANIES PLAYERS

The number of water research centres and enterprises worldwide is impressive but it is difficult to find reliable statistics. Many publications provide an overview of the water market. Several sources estimate the size of the global water market over the last two years has been in the range of $360 billion to $500 billion. According to the president and chief executive of Finnish chemicals firm Kemira; the global water market could be worth $800 billion by 2035. The Japan Bank for International Cooperation established it could reach 100 trillion yen e.g. $1 trillion by 2025.

The Carbon Disclosure project published a report in 2010 about companies engaged or interested by water resources management: "137 financial institutions with assets of over US$16 trillion" backed this report (CDP, 2010). Several websites can provide useful information on the wide range of companies involved in water and wastewater treatment and management:

- The IWA list of Members (company list, corporate list) (http://www.iwahq.org/2w/membership/our-members.html)
- The Platforms of Solutions from the World Water Forum 6 (www.solutionsforwater.org)
- The Exhibition list of major events (Singapore, World Water Forum, Stockholm Water Week, see "Water event")
- The list by Bloomberg of water companies (www.bloomberg.com/markets/companies/water)

Regarding water research, several studies can help to understand the importance of the research sector. At the European level, more than 800 water research projects have received EU support under the Sixth and Seventh Framework Programmes (FP6, FP7). In the first four calls of the FP7 Cooperation Theme "Environment" alone, up to 90 water projects were selected, representing an EU investment of more than €300 million. For the last call of the project, the European Innovation Partnerships received proposals from 100 partner organizations from 20 European countries. In France, the water research community gathers more than 120 research organisations employing more than 4,300 people (ONEMA, Carteau project, 2010). In Germany, the Water Science Alliance established a list of 118 water research institutions.

Worldwide, famous research centres include the AWWA (American Water Works Association, US); the CIWEM (The Chartered Institution of Water and Environmental Management, UK), the IRC (International Water and Sanitation Centre, the Netherlands); IUCN Water Programme (the International Union for Conservation of Nature, Switzerland), the IWMI (the International Water Management Institute, Sri Lanka); the PI (Pacific Research Centre, US), the SIWI (Swedish International Water Institute, Sweden), the CEH (Centre for Ecology and Hydrology, UK), the EAWAG (Research Centre, Switzerland), the Kompetenzzentrum Wasser Berlin GmbH (Germany), the Korea Environment Institute (South Korea), the KWR Water cycle Research Institute (the Netherlands), the CRP Gabriel Lippmann (Public Research Centre, Luxemburg); Centre for water management and research (CWMR, India), etc.

One way to look at the variety of research on water cycle management is to look at the number of prizes and their recipients. There are many prizes to recognize outstanding research work for water issues: the Stockholm Water Prize, the Hassan II Prize, Singapore Innovation Prize, the International Water Association Prize, and the Prince Sultan Bin Abdulaziz International Prize for Water, etc.

The most famous and ancient prize is the Stockholm Water Prize (SWP). When looking at 20 years of rewards, the Stockholm Water Prize recognized a variety of scientific achievements covering the complexity related of water sciences. From the source to the sea, from catchment and back to nature, the history of the award gives an interesting overview at water issues and their priorities throughout the two last decades. We can distinguish two main categories, the work on water cycle, and the work for water management.

Several awards distinguished works on the water cycle and ecosystem as the ones of David W. Schindler (SWP, 1991) on freshwater lakes; on contaminated groundwater from Department of Environmental Engineering at the Technical University of Denmark (SWP, 1992); ecological models of lakes and wetlands by William J. Mitsch et Sven Erik Jørgensen (SWP, 2004). Stephen Carpenter (SWP, 2011) contributed to better understand ecosystem focusing on freshwaters including eutrophication, ecological economics of freshwater, and resilience of social-ecological systems, etc. Hydrology of course, was rewarded in 1997 and 1998. Professor Werner Stumm was distinguished in 1999 for understanding chemical reactions in the water environment, another key issue for water quality management.

For the domestic water cycle, we find: irrigation and water conservation with Madhav Atmaram Chitale (SWP, 1993); wastewater reclamation by Takashi Asano (SWP, 2001) or, an integrative approach to water and wastewater management with Peter Wilderer (SWP, 2003). Wastewater management was recognized with Takeshi Kubo (SWP, 1994) and Perry L. Mc Carty on bioreactors. His works changed the design and operation of water and wastewater systems. Work on health and water was rewarded through the accomplishments of Bindeshwar Pathak in 2009 and Rita R. Colwell in 2010.

It was only in 1995 that water management was recognized and it is all the more revealing that the Stockholm Water Prize awarded WaterAid, a UK NGO. The NGO received the prize for promoting water and sanitation access worldwide. Before, everything was for pure science. In 2000, Kader Asmal (Minister of Water Affairs from South Africa) is recognized for his work for promoting water legislation. In 2005, principles of community participation are acknowledged with Sunita Narain and the Centre for Science and Environment. The concept of virtual water developed by John Anthony Allan is awarded in 2008. In 2012, the conceptual and guiding approach based interaction of man; water management and food, livelihoods and the environment gave the opportunity to the IWMI to gain the prize. Finally, Asit K.

Biswas won in 2006, for its multi-faceted contributions to global water resource issues, including research, education and awareness.

During the Stockholm Water Week, several other prizes are rewarded including a prize for industry, one for younger generations, and one for journalists who have written outstanding work on WASH challenges. The Stockholm Water Prize is part of the many rituals of the water sector and represents important moment to bring cohesion in the water world.

5.4 HUMANITARIAN PLAYERS

Humanitarian players are non-profit associations. There are a large number of non-profit associations working for water and sanitation access, water preservation, and water awareness at the local and national level. The following list provides an overview of non-profit association with an international visibility[1].

Dedicated to water
Charity Water, Water.org, Solidarités International, WaterAid, International Rivers, etc.

With water programs
WWF, Surfrider, Ocean Society, Action against hunger, CARE, etc.

At the national level with international activities and programs:

- Eau & Rivières (France)
- Eau Vive (France & Africa)
- 1001 Fontaines (France)
- Aquasistance (France)
- Eau Secours Afrique (France)
- Watercan (UK)
- Just a drop (UK)
- Join the pipe (Netherlands)
- Waterforpeople.org (USA)
- Watermissions.org (USA)
- Thewaterproject.org (USA)
- www.endwaterpoverty.org (USA)
- Wine into water (USA)
- Drop in the bucket (USA)
- Water without Borders (USA)

5.5 NATIONAL, REGIONAL, LOCAL, AND CITIES PLAYERS

The *Global Water Intelligence* published each year a comprehensive overview of the sector including the survey of more than 5,000 utilities. In Europe, EUREAU established that 10,000 water and wastewater utilities provided water and wastewater services to 400 million EU inhabitants. To list the number of municipalities and regions with dedicated water programs would be rather long. The number of utilities and regions involved in managing water resources actually show the stake of Water Communication. From small to large utilities, each city has an impact in managing water resources. Since they rely on limited financial resources and usually share surface- and underground waters, they are more and more to collaborate and to join into communities of cities. They also get organized at the basin level or in platforms

[1] See the list of 50 top non-profit association and blogs: www.seametrics.com/blog/water-non-profit-blogs

and networks (inter-water). They also are collaborating thanks to the variety of programs developed by funding agencies and the UN.

In the large landscape of actors, several cities are affirming their leadership in managing water resources. Inheriting from discourses on water cities, they focus their communication strategies based on their technological competences or innovation capacities; their offer of new ways of life; or relations with arts. (See Discourse on cities)

5.6 WATER PERSONALITIES

Traditionally, the water sector has counted on voices of many activists, in particular, women. Vandana Shiva has been the first woman to speak up and fight for water and women. She has been joined by many strong water feminine figures including Danielle Mitterrand, Anne le Strat, and Maude Barlow. Famous male militants include Ricardo Petrella who created the International Committee for the World Water Contract in 1993; Marc Laimé, Jean-Luc Touly, or Pablo Sanchez Centellas leading the European initiative "Right for Water". They are opposed to CEOs or institutional figures. The most famous include the CEOs of the two main figures of the two French Water groups, Gérard Mestrallet and Henri Proglio. But others have put their names in the history of this rather French battle including Jérôme Monod (CEO Lyonnaise des Eaux), Antoine Frérot (Veolia), Alain Carignon (Mayor of Grenoble), André Santini (SEDIF), Thierry Vandevelde (Vivendi – group formerly owning Veolia).

In the matter of tap water versus bottled water, several figures emerge: Annie Leonard, Peter Gleick (Pacific Institute), Res Gehringer (Journalist) against the CEOs of major bottled brands: Peter Brabeck (Nestle Water), Pierre Papillaud (Cristalline). At the international level, institutional figures include leaders of networks and associations: Paul Reiter (IWA), Peter Forssman (SIWI), Loic Fauchon (World Water Council), Benitto Braga (World Water Council), Joan Clos (UN Habitat), Gérard Payen (Aquafed), Peter Gammeltoft (European Commission). The King of the Netherlands is famous for his engagement towards water being a chair of the UN General Assembly and Human Rights Council.

Holding institutional responsibilities, many women are also well-known: Catarina de Albuquerque (UN), Sheela Patel (Slum Dwellers International), Blanca Jimenez (UNESCO), Margaret Catley-Carlson (UN), Alice Aureli (UNESCO), Josefina Maestu (UN-Water). Many recipients of the prizes are also famous figures within the water sector as well as experts, as seen before. All those figures are famous for the water sector but rarely seen or occasionally in media and by the general public.

In this regard, among powerful men, Bill Gates gave his support to a new technology for toilets of the future last year. He showed the strong commitment of the Gates Foundation fund development through water access, hygiene and sanitation for all. Richard Branson is also famous for his commitment for water with Charity Water, Water.org and with his own non-profit association, Whole Water World.

Among celebrities, Matt Damon is the most cited star for his engagement with water (Survey on Water Communication, see Chapter 4). He created his non-profit association in 2006. This first association was named "H2O Africa Foundation" (cf. the role of naming H2O in communication). The non-profit association became water.org in 2009 after a merger with another non-profit association named "Water Partners". Since 2009, water.org has been active in many countries and developed several communication campaigns including "One Bottle's Promise", "Happy Toilet Day", and a visit in the fields with Bill Clinton. In 2012, Jodie Foster also participated in a clip about water and women, "For women, it's personal'. But the most famous campaigns are the one with Matt Damon: "True or False" and "Damon Clause" have been very popular. The most famous and recent (explaining the answers on the survey) is the campaign "Strike with me". The campaign was announced through a press conference, followed by two TV spots with three actors (Bono, Richard Branson, Olivia Wilde, Jessica Biel, Jason Bateman and Josh Gad). The campaign on the social media was very popular and gained many supporters (See Box 5.1).

Box 5.1 The campaign "Strike with me" by water.org (Source: water.org, 2013)

Strike with me, a successful campaign

Act 1:
The announcement in a press conference
The "press conference" or "Matt Damon Goes on Strike!" shows Matt Damon talking with journalists in a large press conference. He asks who knows about "Toilet Day" and initiates a conversation with the journalists who seem not to know anything about this event. He provides key figures while journalists react and bring funny perspectives. The announcement "I want go to the bathroom" creates more reactions where the personality and history of Matt Damon create more connivance with the public ("hair", "maths", "Damon's cup"). The dialogue plays with the ridicule of journalists' questions but also allows for more indirect explanations of what it means ("excreting waste from the body, etc. both caca and pipi"). The confusion permits to clear the message: "say no toilets, to say yes to water and sanitation for all". The video has been views by more than 1 million viewers on YouTube.

Act 2:
The support by celebrity in two short TV spots.
#1: The first includes the participation of Bono, Richard Branson and Olivia Wilde. It tells the story of a fight between Bono and Matt Damon. The three actors are telling the stories with references to the personality of Matt Damon (Irish roots) as well as the personality and history of the three celebrities: Bono (Pacifist), Olivia Wilde (Android from the future). It gives key figures to convince the public.
#2: Jason Bateman and Jessica Biel tell complementary stories, one being based on the status of a star (not going to the bathroom, not sweating, developing own brand) and the other on more common – vulgar practices (peeing in the kitchen sink, practical questions). The third character by Josh Gad brings funny interstices by screaming gently, waving his fist.
#3: In both stories, Matt Damon breaks up the story to introduce rhythm but also to reinforce the message. He delivers a key slogan: "If you don't go to the bathroom, you're a celebrity".

Act 3: Social media campaign
The press conference and the spots refer to the website launched by Water.org to engage people: www.strikewith.org

Strike qualified work stoppage or engagement to defend a cause. Famous strikes usually involved unions but many strikes have been led by famous figures, one of the most famous being Gandhi for his hunger strikes. As stated by Goubert (2006), nobody would really consider doing a "drinking water strike". The press conference explained the value of the strike that is symbolic rather than concrete. Medical employees or firemen who go on strike but continuing their activities can compare the strike to strike taken.

The campaign can be found on www.strikewithme.org and Youtube under Water.org

The tradition of supporting humanitarian causes has become a shared public engagement among many celebrities since the early 1950s. The first nomination was entitled to Danny Kaye (entertainer The Danny Kaye Show) in 1954 by the UNICEF. Since the UNICEF needs to get funds from the public, the strategy was to benefit from the recognition of the Goodwill Ambassador to bring messages, to represent the organization and to participate to its activities. They hold a diplomatic mission and can have to write reports, make speeches and to reach to other programs and governments for secular and apolitical missions. The UNICEF considers Peter Ustinov (writer, Goodwill Ambassador nominated in 1968) Audrey Hepburn (actress and Goodwill Ambassador between 1988 and 1989) as their main historical figures. The institution benefits however, from the support of many Goodwill Ambassadors and advocates since 1984.

In 2010, the UNICEF launched the operation "Celebrity Tap". The UNICEF played with its history and tradition of having Goodwill Ambassador to give more visibility to its program "Tap Water Project". Five young celebrities joined a campaign to reach a new audience (See Box 5.2).

Box 5.2 The campaign "Tap Project"/"Celebrity Tap" by the UNICEF (2009-2010)

Celebrity Tap – UNICEF - Campaign 2010

The UNICEF launched a campaign with five young US celebrities reaching different audiences. For each celebrity, UNICEF developed a TV spot using references for the public of the star. On the internet, the design also played with the particularities of each personality.

For Rihanna, the movie is shot in black and white and pictures the singer as "a real international diva". The key values are "class", "elegance" but also play with the sexy figures of the star. Two male voices are speaking, almost whispering and pronouncing key words such as "hot", "mouth". The music is lounge bar and Rihanna wears a dark dress. The movie takes place in a modern type kitchen that conveys the similar ideas of bars, modernity. On the website, the picture of a microphone recalls the singing performances of the star.

For Selena Gomez, the TV commercial plays with the codes of Disney movies. Selena is dressed in a princess dress and speaks to a fictional blue bird coming to her kitchen. In many Disney movies, stories will usually mix drawing with films and have humans talking to animals. Selena obviously lives in a fantasy world that the electrician fixing a light in her living room underlines. From his view, Selena is talking by herself, pouring water to nobody and dressed in her pyjamas. While she talks, she explained the UNICEF Tap project.

You don't have to be a celebrity – UNICEF - 2009

In 2009, it is interesting to remember that UNICEF developed a series of print advertisement using values over "celebrity". Visuals were representing over dressed individuals (mostly women) holding a glass of water. One was styled as a blonde diva or another as a long black-haired singer wearing leather. Both of them were stereotyped representations of stars but also reminiscent of famous US singers.

Box 5.2 Continued

The slogan "You don't have to be a celebrity to make a difference" put the emphasis on the individual act and was a way to make fun of the person obviously a mere reproduction of famous people. The 2009 campaign plays on the value of individual whereas 2010 seek for endorsement through famous figures.

Today, the project is supported by: Alyssa Milano (actress), Angie Harmon (actress), Bridgit Mendler (actress), Common (rapper), Emmy Rossum (actress), Heidi Klum (model), Jeff Bridges (actor), Judd Apatow (director), Kat Graham (actress), Kate Walsh (actress), Kevin Spacey (actor), Kristen Bell (actress), Marcus Samuelsson (restaurant owner), Nas (rapper), Seth Rogen (actor), Sting (singer) and Lenny Kravitz (singer) have all supported the WASH programme since 2012.

Many other non-profit associations are using the strategy of endorsement by a celebrity to further support their cause. The founder of Charity Water, Scott Harrison was formerly in Public Relations, organizing events and working with many celebrities. He is famous for organizing events with celebrities attending. Several have contributed to the communication campaigns of Charity Water. Several participated in a short TV spot (Jennifer Connelly), organized events to raise money or some participated in the "Birthday challenge" and participated in the construction of wells (Will and Jada Smith). Among other celebrities, Charity Water is supported by Adam Lambert (American Idol), Adrian Grenier (actor, he also supported UNICEF Tap Project), David Blaine (magician), Jenna Elfman (actress), Mathew Modine (actor), Terry George (writer), Richard Branson (businessman), Yaya Da Costa (actress) etc.

WaterAid, the UK based non-profit association is supported by several British celebrities including Miranda Hart (actress), Christine Bleakley (TV presenter), JLS (Band), Ringo Starr (musician), Tommy Walsh (TV actor), Tamzin Outhwaite (TV actress), Adam Hart-Davis (TV presenter), Cyrille Regis (Footballer), Helen Lederer (comedian), Rachel Stevens (singer), etc.

The non-profit association "Whole World Water" supported by celebrities but from very different sectors and not all from show business: David de Rothschild (activist and adventurer), Shekhar Kaipur (film maker), Stanley Johnson (environmental expert), John Hitchcox (Design company), Graham Hill (entrepreneur), Ben Elliot (founder of Quintessentially), Rob Collins (CMO of Tsogo Sun), Yves Béhar (Designer).

In France, the non-profit associations are usually not supported by public figures. Many well-known faces for the French public however, joined the cause of the militant non-profit associations. The Foundation France Libertés obtained the support of Yann Artus Bertrand (Photographer), José Bové (Militant), Peter Beal (Producer), Bertrand Delanoë (Paris Mayor), Anne Le Strat (Deputy), and Titouan Lamazou (Navigator). The initiative "Right for water" gained the support of Marc Levi (Writer). During the World Water Forum in Marseille, the TV presenter and producer Bertrand de Villardière gave his support and produced a movie for Action Against Hunger. Yann Artus Bertand presented his second "Home" called "Water in sharing". The movie was co produced by the French Development Agency. The support and attendance of Famke Janssen (actress) at the World Water Forum was also a new initiative to bring more visibility to such an institutional event.

Needless to say, there are enough supporters of the water sector. Nonetheless, water professionals are most likely convinced that few public figures exist. The survey and interviews showed that few professionals were aware of such important support by celebrities. It argues for the lack of key visible expert or professionals in the media. Many professionals feel there is no strong advocate both famous and expert bringing visibility on water to institutions, governments and other stakeholders. For instance, many argued that Al Gore really brought the case of climate change to a higher degree of visibility and knowledge from the public. Who could be the public figures of the water sector? Actually, they are many young figures engaged with water and known in their countries. For instance, in France and Canada, two pairs of adventurous brothers are hosting a TV show about water (see Box 5.3). Their documentaries are original, creative and propose scientific content with branded adventure and feeling of travel that enable the public to learn while enjoying their discoveries, etc. A receipt for new pedagogic approach to water, isn't it?

Box 5.3 TV broadcasting, Pedagogical documentaries by adventurers' figures in France and Canada

The Water Brothers

Who?
Alex and Tyler Miflin
What?
Series of 26 minutes documentary
Where? When?
On TVO, Canadian Channel

The concept:
Alex and Tyler develop one key challenge per episode. For each episode, they carefully name and brand the content: Watergate, No woman, no water, Carpageddon, etc. They address key issues in a friendly accessible tone and share their adventure throughout the world.

Their website is simple, well done combining educational information, library resources and even Quench, a phone application that helps to find fountains and public water points in the city of Toronto.
More: http://thewaterbrothers.ca/

Les Frères la Tullaye

Who?
Geoffroy and Loïc de la Tullaye
What?
Series of 26 minutes documentary
Where? When?
On France 5, French Channel

The concept:
Loïc and Geoffroy developed series of episodes around an adventurous path built along the mythical silk road or the great rift valley. For each expedition, they investigate one key challenges to water management such as the concept of virtual water, the economic development of the river, or food and water. Their approach is complemented with conferences in schools and in companies. They initiated a blog for their last expedition and continue to provide updated information about world water initiatives.
More: www.lesfrereslatullaye.fr

5.7 WATER EVENTS

The water sector meets regularly to exchange, discuss and conclude many business contracts. These rituals are numerous and counting the number of water events worldwide is a difficult task. Each year, the UN invites individuals and organisations to subscribe to their website in order to list the number of activities undertaken for the World Water Day every year. Other sites also provide information on water events serving specific public such as INBO, the European Water News, etc. Looking on Internet, the research of "Water Week" brings 1,810,000 results in 18 seconds. There are Water Weeks at the global, regional, national and local levels (See Box 5.4). They are organized by major water organisations, professional event planners or non-profit organizations.

In the water sector, there are four major meetings:

Box 5.4 Examples of Water Week in the World

International

- World Water Week in Stockholm (1991), www.worldwaterweek.org
- Singapore International Water Week (2008): www.siww.com.sg
- International Water Week in Amsterdam (2011): www.internationalwaterweek.com

Regional

- Water Week Latino America (2013): www.waterweekla.com
- Africa Water Week (2008): www.africawaterweek.com
- Asia Water Week (2013): www.adb.org

National

- Canada Water Week (2013): http://canadawaterweek.com
- National Water Week by the Australian Water Association (1993): www.awa.asn.au/nationalwaterweek
- India Water Week (2012): www.indiawaterweek.in
- Drinking Water Week (1988): www.awwa.org

Local

- Water Week in UK organized by OXFAM: www.oxfam.org.uk
- Vancouver Drinking Water Week (2013): www.drinkingwaterweek.org - BC Water and Sanitation Association – Vancouver / Canada
- Salt Lake City Water Week (2013): www.slcgov.com/waterweek

In brackets, the year of creation of the Water Week.

Most of the Water Weeks are recent. It shows the recent increase of the phenomenon. The American Water Works Association has organized the longest-serving water week since 1988. The most famous in the water sector and probably in the media as well, is the Stockholm Water Week.

1 The World Water Forum organized every three years by the World Water Council.
2 The political week of water in Stockholm, which punctuates the summer each year.
3 The half-conference, half-summit, half-exhibition and "business oriented" International World Water Congress is organized by the IWA every two years.
4 And finally, the technological oriented SIWW, the Singapore International Water Week organized in June every year.

Many question the objectives and legitimacy for such big events. The organisation of such events is costly and oriented by the political agenda rather than needs and strategic developments of the water sector. The organisation of events however participated to the construction of a water community, to connect professional from a highly fragmented sector, and to convey a sense of vision and shared values.

6 Inter-National level Setting the agenda

According to Gobin and Deroubaix (2012), the analysis of discourses by international organisations is still a relatively unexplored research area. Della Faille (2011) states that the study of discourse was introduced during the 1990s for international development studies. His mapping shows the emergence of contributions on many different topics (gender, power in development, and most famously, on speech and politeness). He also underlined that "these authors remain marginalized in their own field of study and their work ought to be circulated in general discourse analysis circles." (Della Faille, 2011: 215) Several are focusing on the analysis of the discourse from international organizations including international governmental agencies and non-governmental organizations. The main common trait is the criticism of the "mainstream analysis of underdevelopment" (Della Faille, 2011: 215) and that "international development as it is practiced by government agencies and private organizations has not internalized these criticisms" (2011: 232). References provided by different authors will usually date from the late 1990s (Dontcheva-Navratilova, 2004; Das, 2008; Lehtonen, 2009).

The discourse of international organizations is observed from a power relation quoting in particular the work of Foucault (1977) and Gamsci (1971 see also Robbins, 2003; Walker, 2006; Chacon-Zuloga, 2011; Della Faille, 2011; Mukhtarov *et al.*, 2012 and Goldman, 2005). Relations of power will usually point out the supremacy of the non-governmental and non-profit organizations in providing the same type of vision on development and "underdeveloped countries". This supremacy is today questioned by the multiplication and raised visibility of the scientific work on discourse analysis of international organizations (Gobin *et al.*, 2010).

The analysis of discourse organizations at an international level is a complex approach. It can be characterized by the multiplicity of organizations producing discourses, by the multiple interactions from these organizations and the multiple ways and tools conveying their discourses. Historically, discourse analysis has focussed on contexts and interactions rather than meanings (Berger and Luckmann, 1967). Discourse analysis is based on a heritage from linguistics, narrative structure (Propp, 1928 trad. 1965; Greimas, 1966) and public sphere interactions including activities of speech, rhetoric and discourses. The approach by Habermas (1962 trans. 1988) of the public sphere, involves these interactions by actors, both individuals and organizations. The construction of such public sphere is based on the search and building of visibility built by oneself (Goffman, 1959), by the apparition on the public sphere and what one's wants to publicize (Arendt, 1983) and based on interactions between organizations, the media and context (Veron, 1997; Charaudeau, 1997). Activities of constructing discourses are founded on speeches, the abilities to take the floor and to address the public(s). They are the results of a context, social interactions, processes and the performance of verbal and non-verbal communication.

In this frame, discourse analysis is a methodology that understands and reflects the product of interactions, processes, dynamics between actors serving the construction of a reality (Berger and Luckmann, 1967), a public sphere (Habermas, 1962), and international codes and norms (Finnemore, 1996). It constructs the agenda from a constructivist perspective based on discourses from actors and power relations (Fierke, 2002 and 2005) According to Finnemore, "international organizations pick up on norms or concerns and popularize them" (quoted by Das, 2008); they are "platforms" promoting discourses (Finnemore and Sikkink, 1998) and have a strong ability to set priorities on the political agenda, on national powers or individuals diffusing norms, rules and ways to understand concepts and the world (Barnett and Finnemore, 1999 and 2004).

It is important to underline the work of Hajer (1995) linking environmental discourse and global organisations. Hajer is considered as one of the first to study links between policy, ecological values and discourses. Here one the first to analyse impacts from governments on environmental politics and policy. His framework analysis of discourse points out the paradox of policy discourses on environment: governments are promoting a shift for the economic model and new shared democratic process of setting new economic objectives but rarely undertake global changes. This dynamic will bring criticism including insufficient actions that will actually be convenient to people adopting discourses that will serve their interests (Hajer, 1995: 261).

In the water sector, the study of global discourse from non-governmental organizations and non-profit organizations, from global private companies, international water networks, conferences and major events declarations are recent and also limited. Originated from the widespread of social approach in water management, analysis of water discourses can be found in many local case studies. At the global level, some first scientific work considered the impacts of water politics on local territories (Feitelson, 2002); climatic events and meteorology (Van Beek, 1993; Jairath, 2003); privatization (Robbins, 2003) and commodation (Bond, 2004); IWRM (Allan, 2003); water wars and justice (Tarter, 2005) and trans boundary basins (Mukhatarov *et al.*, 2013).

The main contributors are A. Biswas (2004 and 2009) and Mount and Bielak (2011). Other interesting publications related to global discourse on water can be found on global organisations such at the UNESCO (Dontcheva-Navratilova, 2004; Tavani, 2013), the OECD (Lehtonen, 2009), etc. Several publications from Masters thesis and doctorates are screening specific aspects of water discourses by the UN Galaxy and major funding organisations. There are focusing on specific topics and / or geographical scales.

Based on those elements, water global discourses and global processes producing globalized discourses are based on linguistic processes, relations to power, environmental discourses and policy providing norms, rules and framework to individuals setting a globalized discourse to see and understand human relationships with water. Water discourses are characterised by a paradox promoting water as a "vital resources" unifying all humans at the global level and for a universal cause, but without allowing long-term changes thanks to the multiplicities of visible discourses allowing people's, governments, organisations and individuals to interpret and produce discourses institutionalizing or conceptualizing their own interests. This plasticity or "liquidity" (Mount and Bielak, 2011) of discourses constrains promised changes into active actions. We further develop this analysis from (1), an historical overview of major UN conferences and resolutions where the liquid conceptualization of water challenges appeared (2) the framing of universal campaigns to unify actions on a globalized water vision based on the analysis of the UN Water Decade campaigns in particular "Water for all"; (3) the analysis of a specific topic showing the building an international topoi serving this unifying process: water and women.

The institutionalization of water discourses originated from first conferences on water management involving several nations. Looking at the history, first references can be found to treaties to share water resources, watersheds or transboundary rivers (Sironneau, 2002). Looking at the history of laws, treaties and legal actions on water resources, the most common quoted references initiating the sharing of water resources or organizing water management are related to hydroelectricity and waterways (Hassan, 2010). It is only with the organization of the UN conference in Mar del Plata that other distinctive discursive elements emerged when talking about water (Biswas, 2004). This recognition originated from the UN resolutions, conferences and other organisations setting new approach to water management such as the European Union, the UNEP, etc. The slow process of recognizing the importance water on the public sphere is however limited by paradox of the global discourses and global agenda setting. The "liquidity" of texts, wordings, and topics (Mount and Bielak, 2011) along with speech characteristics and formats

(Dontcheva-Navratilova, 2004) are putting limits to implement concrete changes at the local scale. These limits are similar to the ones identified by Hajer (1995) on "environmental discourse" and confirming the over confidence put into the communicational potential of environmental discourses bringing distrust and disputes (Libaert, 1992).

6.1 HERITAGE OF INTERNATIONAL WATER CONVENTIONS AND CUSTOMS (805/1921–1972)

Sironneau (2002) reminds that international conventions are referenced and refers to Charlemagne's period for the appearance of a first agreement on the usage of Rhine River crossing several regions (805). Sironneau established that more than 3,800 acts have been signed between 805 until 1983. These conventions are setting the sources of the international regulation and water rights. In 1815, the Vienna Treaty is the first step towards the international use of European rivers; this treaty is usually quoted as a reference for being one of the first instituting transboundary management and uses.

The establishment of an "international custom over water uses and sharing" corresponds to the setting of a glossary, concept and principles towards water management. Sironneau enounced the following characteristics as part of the establishment of the right to water:

- *To collaborate and negotiate in order to reach an agreement*
- *To build infrastructures that won't impact other States*
- *To consult other States before any intervention*
- *To divide and use equally shared resources including groundwater*

The acceleration of economic development and the need to manage water resources slowly changed the establishment of water management, water science and technology and consequently, the setting of international conventions. The building of dams and reservoirs particularly impacted water management from a local and basin river oriented management to a more global scale (Hassan, 2011).

The main reference for setting an international regime for waterways dates from 1921. This convention established in Barcelona on April 20th 1921 is organized by the League of Nations (LN). As such, it shows the influence of the first international organisations created by States after the First World War. On December 1923, the LN further ruled on hydraulic forces and their infrastructures in Geneva. These two first international conventions are showing how water issues were perceived in the early 20th century. Water is only considered as a natural resource that can be use for transportation and economic purposes. Discourses on water will only concern the sharing of resources from which States might extract added value for their development. This approach will be the main and major way of structuring right to water over the next 50 years until 1977 and the UN Conference in Mar del Plata; the UN resolutions voted by the General Assembly can help to trace the setting of such references, norms and discourses on water challenges.

It is important to remind that one of the first resolutions from the UN is related to the declaration of rights and duties from the States. Since its beginning, the UN has wanted to act and organize the world managing two opposed forces: setting resolutions overruling States under the principle of "duties" or mandatory principles that will serve superior ideals and the organization of the global agenda; as opposed to the respect of national sovereignty and the rights of States to organize their territories, people and legal systems.

Through the examination of UN Resolutions (22 important resolutions related to water including resolutions on natural resources, biological resources, climate or any texts that might refer to water resources (see Box 6.1), resolutions on global challenges appear with the organization of the UN as a global institution following the vote of the Universal Declaration of Human Rights on December 10th 1948.

Box 6.1 UN Resolutions on water resources between 1946 and 1977

1 **1949/374 (IV):** during the fourth session, the UN recognised for the first time, the priorities of setting legislations on territorial waters and under international laws.

2 **1952/626 (VII):** the UN recognised the right to exploit natural resources and particularly forbids any interventions from Member States that could prevent any other Member States from enforcing its sovereignty over its natural resources.

3 **1954/900 (IX):** the UN organised the first international conference on biological resources from the sea.

4 **1957/1105 (XI) and 1958/1307 (XIII):** the UN organized the first international conference on the law of the sea.

5 **1958/1306 (XIII):** Launch of a study on historical waters, in particular historical bays.

6 **1959/1401 (XIV):** Launch of a study on legal aspects related to the use of international rivers.

7 **1962/1831 (XVII):** The UN recognised the importance of planning and anticipating the preservation of natural resources regarding the economic development. It is interesting to note the inner paradox set by the UN throughout time. Whereas the national sovereignty is confirmed and promoted several times (including during the 17th Session, Resolution 1897), this resolution aims to promote conservation measures for natural resources. One of the main principles wants to ensure the preservation of natural resources revolves in presenting its utilities to people. It means integrating the preservation of natural resources into national strategic plans and legal systems… States are encouraged to take the responsibility for ensuring the implementation of such systems. One concrete recommended action is promoting public's awareness:

> "(f) Organizing national campaigns through educational institutions and through the press, radio, television and all other possible media for the dissemination of information in order to secure the co-operation of the population in the achievement of these aims."

The paradox seems to appear between respecting Member States (who want act to preserve natural resources) but encouraging them (and potentially initiating awareness campaigns rather than radical changes within the legal system to protect natural resources).

8 **1965/2093 (XX):** First resolution on natural catastrophe.

9 **1966/2171 and 2172 (XXI):** Launch of studies on Natural resources (evaluation) and resources of the sea.

10 **1968/2386 (XXIII):** Apparition of the concept of human environment where human activities can pollute and impact the environment including water quality.

11 **1969/2569 (XXIV):** Recognition that countries without sea coast may have more difficulties to access to economic development

Box 6.1 Continued

12 **1969/2580 (XXIV):** Resolution for coordination over marine activities.

13 **1970/2669 (XXV):** Introduction of the codification of the rules of international law relating to international watercourses.

14 **1970/2750 (XXV):** Reservation of resources from the sea and the ocean "in the interests of mankind"

15 **1970/2657:** Recommendations for the organisation of the UN first conference on Environment.

16 **1971/2849:** The resolution recognised links between "Development and Environment" recognising that "development plans should be compatible with a sound ecology and that adequate environmental conditions can best be ensured by the promotion of development, both at the national and international levels" and to further promote joint environmental policies and development policies.

17 **1972/2996:** The General Assembly in short three paragraphs that resolutions adopted during the 27th Session won't affect the Principles 21 and 22 voted during the UNEP. This resolution is recognising the "International responsibility of States in regard to the environment".

Principle 21

States have, in accordance with the Charter of the United Nations and the principles of international law, the sovereign right to exploit their own resources pursuant to their own environmental policies, and the responsibility to ensure that activities within their jurisdiction or control do not cause damage to the environment of other States or of areas beyond the limits of national jurisdiction.

Principle 22

States shall cooperate to develop further the international law regarding liability and compensation for the victims of pollution and other environmental damage caused by activities within the jurisdiction or control of such States to areas beyond their jurisdiction.

18 **1974/3244:** The UN acts on assisting Bangladesh after a severe flooding, a first active resolution to assist a country after a natural disaster.

19 **1974/3264:** New step forward to protect the environment and the climate from military purposes directly implying to Member States that nuclear weapons tests should be stopped. This is the first time "climate" appears.

20 **1974/3337:** The UN rules for an international cooperation on Desertification, the term is appearing for the first time in the title of the UN resolutions.

21 **1975/3513:** The UN welcomes the organisation of the UN Water Conference in Mar del Plata.

22 **1977/A/RES/32/158:** The UN confirms the action plan adopted in Mar del Plata and requests the implementation of plans and programmes at the national and regional levels.

The appearance of environmental concerned is clearly accelerated at the early 70s but it is interesting to note that many texts written from 1949 until 1972, are setting many key principles and statements that are later repeated and used as important guiding concepts and principles. The organization of the conference on the environment is the occasion of many resolutions related to environment.

Starting the organization of first international decades, the increasing concerns over the uses of natural resources and questions on economic development models, the discourses on water are "broadening". When water, natural resources or climate impacts are mentioned, the resolutions introduce new terms, new challenges, new groups of individuals and most of all, new concept for economic development. This shift of the promotion of development and uses of resources is clearly introduced in 1974 with the definition of a "New Economic Order".

6.2 A PARADIGM SHIFT, A DISCOURSE FOR GLOBALISATION (1974–1992)

> "Solemnly proclaim our united determination to work urgently for the Establishment of a New International Economic Order based on equity, sovereign equality, interdependence, common interest and cooperation among all States, irrespective of their economic and social systems which shall correct inequalities and redress existing injustices, make it possible to eliminate the widening gap between the developed and the developing countries and ensure steadily accelerating economic and social development and peace and justice for present and future generations." (United Nations, 1974)

The importance of language and the role of UN declarations, statements and large conferences have been underlined in setting the global political agenda and visibility of important world's challenges (Gobin in Deroubaix: 2010, 107). Biswas (2009) analysed such an influence on the particular issue of water management). The specific rhetoric and language set up by the UN slowly conveyed the roots and definition of new economic models as shown by the first resolution by the UN for a new economic order.

The UN resolutions are included and repeated by major conferences starting with the UN "megaconferences" (Biswas, 2009). In those circumstances, the role of major conferences in setting new paradigms is also a key component to setting new discourses to change practices and initiate programs at the international scale. They are "founding texts" acting as undisputable reference and setting key definitions (Espuny-Pascual, 2007). When looking at their evolutions of thematic, it appears that environment takes a more and more quantitative importance, but even this importance remains relative and limited compared to other important issues.

The lexicography densified and many terms appeared over time: water (1949); natural resources (1952); biological resources from the sea (1954); historical bays and waters (1959); nature (1962); oceanographic activities (1969); environment (1970); links between development and environment (1971); water – organisation of conference (1975); marine pollution (1979); future generations (1980); ozone layer (1987); climate protection (1990) and Sustainable Development in 1991. These terms are accompanied by a multiplication of terms showing an acceleration, repetition and multiplication of official resolutions related to environment or development. Terms participate in the writing of the institutionalization of water references, codes and dynamics within the professionals, experts, and governments, etc. Any stakeholders involved in water management. The UN conferences seem to be aiming at gathering people to help the identification of common knowledge and practices, it clearly wants to institute networks, thematic groupings, information exchange and interactions. As such, the action of the UN served the objective of building processes around a common objective for joint action. The analysis of the UN Resolutions clearly shows a strong involvement for the Earth Summit in 1972 followed by a strong decrease until 1987. The Brundtland report in 1987 initiated a new political interest for environment until 1992 (See Figure 6.1).

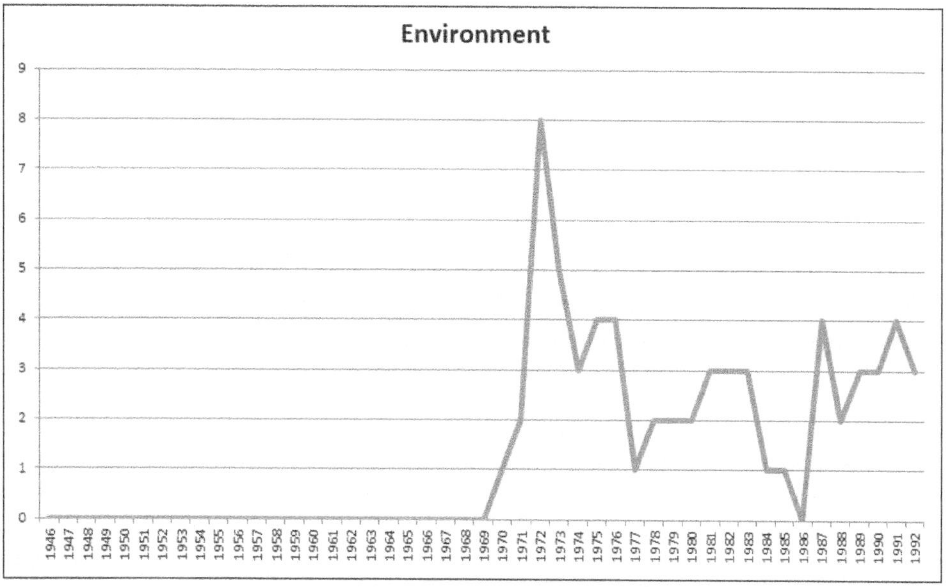

Figure 6.1 Evolution of the topic "Environment" in the UN Resolutions 1946-1992.

Similarly, from 1977 to the early 1990s, the theme of water is marked by an almost complete disappearance of the international arena despite the decade and sanitation (1981-1990). Starting 1988, water is relatively more mentioned in the titles of the UN resolutions (See Figure 6.2).

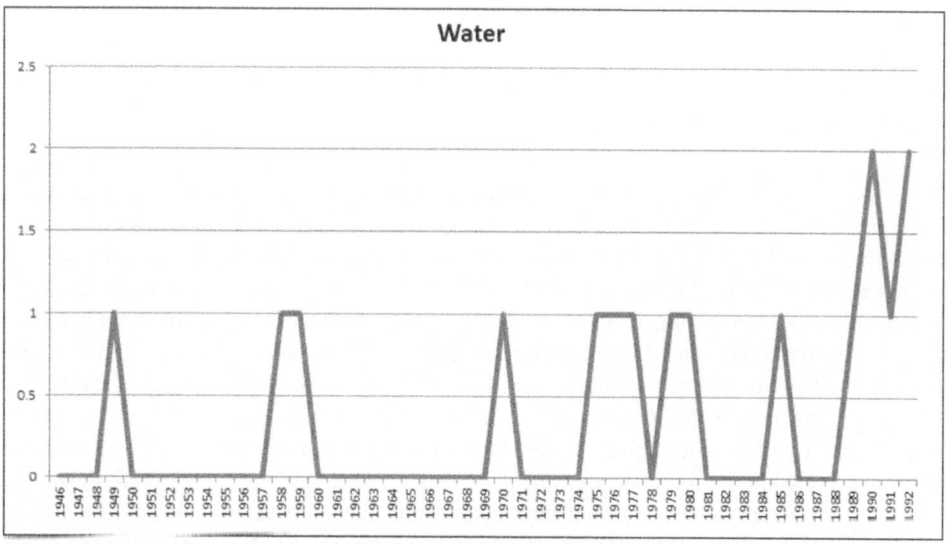

Figure 6.2 Evolution of the topic "Water" in the UN Resolutions 1946-1992.

The acceleration of number of Resolutions in 1988 originated from the end of the UN Decade and conclusions on the activities undertaken during the Decade. There is also a growing concern about water resources in preparation of the Earth Summit and the preparatory conference in Dublin. Meanwhile, at the international public sphere, there is a clear acceleration of the numbers of conferences on water after 1992 (see Annex 2). The third Earth Summit in Rio represents for many experts and for the international community, a new era in the cycle of international conferences and political engagement for environment, Sustainable Development and water resources management. This engagement marked an acceleration and new visibility of water due to the increasing number of events, the creations of new organisations and the framing of water issues in the media. This visibility is however not reflected into the UN resolutions where the "conceptual liquidity" over water management and its priorities seem to limit communication and political engagement.

6.3 FAILURE TO LAUNCH, THE INNER LIMITS OF CONCEPTUAL "LIQUIDITY" (1992-2010/15)

In 1992, the International Conference on Water and the Environment called the "Dublin Conference" defined four pillars to water resources management. Looking at the text, it appeared as a confirmation of the principles defined in 1977 during the United Nations Water Conference in Mar del Plata. The four principles set out in Dublin however acted as founding principles. They are largely repeated in conferences, declarations and appeared to be a key reference to the water community. It played the similar role of the guiding definition by the Brundtland Report on "Sustainable Development". The four pillars of the Dublin Conference acted as a consensus between experts in order to disseminate those pillars to national and international programs and legislations.

- *Principle No. 1 - Fresh water is a finite and vulnerable resource, essential to sustain life, development and the environment*
- *Principle No. 2 - Water development and management should be based on a participatory approach, involving users, planners and policy-makers at all levels*
- *Principle No. 3 - Women play a central part in the provision, management and safeguarding of water*
- *Principle No. 4 - Water has an economic value in all its competing uses and should be recognized as an economic good*

The Dublin Principles
Dublin Statement on Water and Environment, 1992

One of the main goals of defining and proclaiming four principles to water resources management was to influence the Earth Summit in Rio. As such, Dublin is considered as a failure. The Rio Declaration did not include any mention of water management principles in its final declaration. However, it included a Chapter on water resources in the Agenda 18 where water is recognized as a natural resource and its preservation is encouraged. The Dublin conference failed to bring more inputs to the international agenda.

Starting 1992, the number of organisations and conferences accelerated with the creation of the World Water Council (1996), the regular World Water Forum (starting, 1997), the many new networks and professional associations (for instance, the IWA was created in 1999). The multiplicity eventually contributed to a certain reproduction of programs, actions and even, repetition of discourses and solutions. Biswas observed such limits in 2004 after the World Water Forum in Kyoto, questioning the efficiency of such meetings and international statements:

> "The water management profession is now facing a problem, the magnitude, complexity and importance of which no earlier generation has had to face. In the run-up to the 21st century, our profession really has two choices: to carry on as before with a "business as usual" attitude that attempts to solve future complex problems on the basis of experiences from simpler problems of the past, or continue in earnest an accelerated and truly genuine effort to identify the real problems of the future and face the overwhelming challenges collectively and squarely by implementing workable solutions within the short timetable available to us. One of the main lessons of the past has to be that the time for rhetoric's and using one-minute sound bites are over. We must develop urgently new and cogent solutions, which could be operationalized in developing countries and in the fields. Conceptual attractiveness alone is no longer adequate." (Biswas, 2004)

Mount and Bielak issued a major contribution on global UN discourses on water conferences in 2011. They argue for the "liquid" conceptualization over water management provided by UN discourses showing their thematic repartition and changes in setting these priorities (See Figure 6.3).

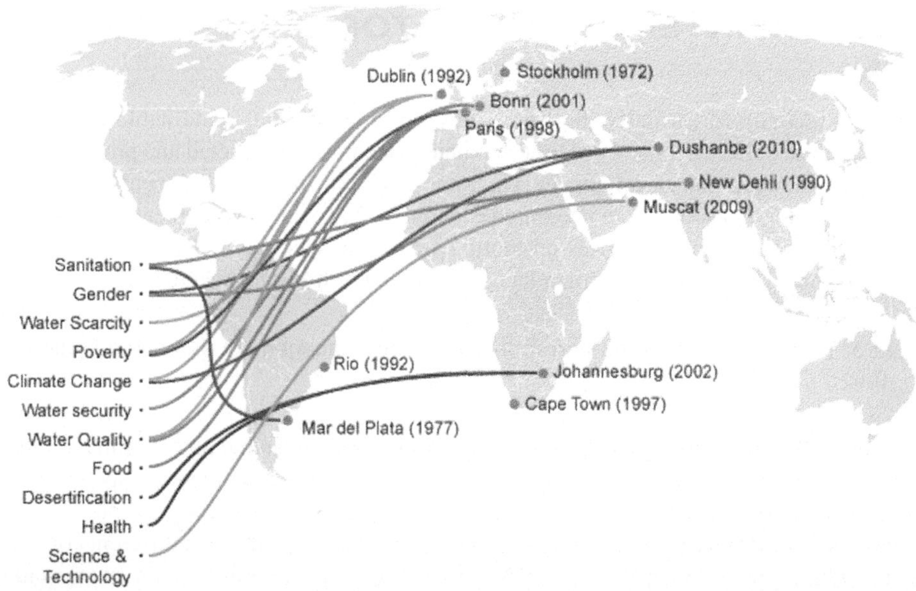

Figure 6.3 Thematic priorities set by the UN conferences over water management (*Source*: Mount and Bielak, 2011).

This liquidity and multiplicity questioned the capacity to bring changes and to influence national agenda for the implementation of stronger water policies. At the end of the World Water Forum in Marseille, many questioned the impacts of such a major conference. The same issue was raised after the deceptive Rio+20 meeting. The effort put into organizing those major events usually disappoints the media and the public who don't grasp the impacts of slow words and diplomatic processes. It questions the communication strategy of those events towards the media and the public but also, towards policy makers and political deciders.

On March 5th 2012, the UN issued social media updates and a press release announcing that MDGs for water access would be reached by 2015. The communication by the UN shocked many professionals who didn't understand why the UN made such an announcement only a few days before the start of the World

Water Forum. Many blamed the UN for bringing such news that undermines the challenges of providing water access before the meeting of high level policy makers and deciders in Marseille. Others also blamed the inexactitude of the data and the media chose to use the data provided by the non-profit association Solidarités International.

On one side, the UN wanted to promote positive news and bring visibility on efforts to achieve better access to drinking water. On the other side, the communication raised scepticism among the public as seen on the UN Facebook page: many brought testimonies from their countries stating water access according to the UN standards was far from being "access to clean water". Other discussed the intention of publishing such news before the World Water Forum asking if the UN wanted to undermine the meeting. Today, the UN has made an effort to review their position and better communicated on details of the definition and criteria to access to clean water taking the criteria set by the WHO for whom the water source has to be within 1,000 metres of the home and collection time should not exceed 30 minutes. Such reactions and changes of strategies questioned the capacity to communicate on the international political agenda, the institutional communication by the UN and the communication towards the public, and how to mediate and inform target groups about water challenges.

6.4 BUILDING INTERNATIONAL TOPOI: WATER AND WOMEN*

In 1991, the UN recognised formally the links between environment, sustainable development and women (United Nations, 1991). Since 1974, international discourses constructed and promoted the need to integrate women to the economic development. The "new economic order" (UN encouraged the promotion of equity, access to natural resources, an integrated economic and social development. In 1992, the Dublin conference evicted four guiding principles promoting water management and sustainable development. The third principle identified the crucial role played by women in water management.

> "Principle 3: Women play a central part in the provision, management and safeguarding of water." (WMO, 1992)

Often tasked with household management, subsistence farming and caretaking of family health, women and children are the primary gatherers of water. As the literature shows, women can spend up to 5 hours daily and walk from 5 to 15 km per day collecting water. Ever since the Dublin principles were declared, organizations have been promoting women's roles in water management through different communicatory approaches including gender, Since the early 1980s, a standard of communication on "water and women" around one key representation has emerged: the traditional role of women walking to get water at the well.

While women were identified as important players in water management in 1992, links between women and the environment were already accepted in major world conferences as far back as 1985, during the third United Nations Women's Conference in Nairobi. Over time (between 1992 to 2012), international conferences slowly changed their use of words and terms quoting: "traditional role played by women", "competence", "participation", "rights", "decision making", "empowerment", "representativeness", and finally, "equality". The interaction between words, concept and actions followed "attempts to address gender issues" (Mount and Bielak, 2011: V). The slow evolution of statements is "strikingly strong" and offers "some of the strongest, most robust language on gender issues of all the Declarations under study" (Mount and Bielak, 2011: 22). By this argument, the authors are reinforcing our observation of this similar and repetitive construction on the issue of involving women in water management (Hervé-Bazin, 2012).

*Part of this chapter has been published in WH2O Journal, n 3 in January 2014, www.wh2ojournal.com

Over the last 20 years, the woman at a water pump is a predominant image in communication campaigns designed by organizations from the UN. As representations tend not to evolve throughout time and among different brochures, this image of the women at the pump has become the iconic face of the women and water rights movement.

Figure 6.4 A historical construction: women walking and carrying water. Over time, the representation of women remains the one of fetching water by feet. (*Source:* from left to right: IRC (1985), UN Water & Decade (2005), UN water & World Water Day 2013, and UNIFEM for the World Women Day in 2013.).

The visual repetition shows how water and women are linked throughout a simple representation. This image brings the viewer to a symbolic heritage of a woman at the well. It reminds us that women are water managers for a wide-range of cultural and religious practices including maternity, marriage, social life, health, gardening, social values, purity and rituals.

At the local ground, women are the first mediators; they are the first targets of communication campaigns and focus groups. Social animators in charge of the program will primarily engage a dialogue with women. When confronting the results from local actions with the communication at a global scale, there is a consistency between reality and depicted images that don't necessarily ease the process of addressing gender gap.

This gap underlines a paradox between representing one stable image of women carrying water and how to change such an inequity in reality. Women are seen as traditional and passive; it appears difficult to frame them another way both in image and discourse.

The rhetoric on water and women reveals the specificity of perceptions to this challenge brought by the central figure of women at the well. We identified some characteristics accompanying this image of women fetching water such as poverty, tradition, and universality, etc.

Poverty: dwelling or unhealthy livelihoods

To depict poor conditions, women are usually poorly dressed, pictured in a simple and sparce environment (housing, natural trails, dirty environment). These images tend to symbolize the vulnerability of the group.

Tradition: low-revenue and economic activities

Women are working in the field (crops plantation, aquaculture, fisheries) or selling agricultural items (crafts, hand-made products). Their activities are usually low revenue activities, awarding women a small income to buy vital goods for their family. Pictures convey a vision of active women continuing activities based on their traditional gender roles.

Universality: the diversity of faces and cultures

The legitimacy of the discourse is based on its globalism and universality. The particularity of the discourse is to address two essential challenges: "water", a resource for life; and "women", half of humanity. Based on this fundamental human imperative, the rhetoric is established to promote a cause bounding people worldwide. Pictures underline the diversity of concerned persons where women play a major role.

The choice made by the editors of these publications is to picture the daily life of women through these images. The usual corresponding discourse is pro-change rhetoric that looks for a radical shift in gender roles.

The building of this advocacy resulted in the definition of fixed image based on tradition, symbolism and a reality which is difficult to change, constituting unique topoi for communication processes. The observation of communication campaigns points out a non-visible conception of water management related to women. This non-visible conception can be observed thanks to analysis of discourse or when comparing communication supports. Whatever the type of organization (association, public entity, private company), there is a certain idea of communicating water built on a perception of customers' expectations to see "water and women" and their traditional roles on the ground. This is because we humans tend to distil cultural diversity into images depicting "one" human being achieving "one" task, i.e., women fetching water at the well. This image corresponds to topoi, a common and shared discourse that people will share in consensus (Guilbert, 2008).

In terms of communication sciences, it has built the process to its "communicative chore". It has eased its potential for dissemination, cultural and political adaptation throughout its visual and rhetoric construction. Based on the TOPOI model (Hoffman, 1999), representations of women in water management are built on:

1. Universality: the image of water and women is universal, conveying the value of life.
2. Interaction: the process focuses on interaction rather than culture. The different tools promote training, education, and participation to avoid cultural disputes.
3. Circularity: the communication is based on a link between funding agencies and the UN (international scale) with funded programs and women (local scale). The process is based on a top/down communication.
4. Reflectivity: the process is built on feedback and shared opinions adapting to the political will, societal change or economic context inflecting discourses and images to adapt to the public sphere.
5. Framing: the discourse is alarming, pessimistic based on the vulnerability of women but promoting faith in people. The communication is negative but women and humanity bring optimism in general (Hoffman, 2002 and Hoffman, 1999).

The process of communicating on the role of women in water management constituted a unique phenomenon of summarization in order to face the major challenge of involving women in water management. The discourse wanted to simplify, multiply and mainstream the advocacy suggesting profound social changes.

If this rhetoric has brought positive changes globally, at a local level, it might have simply crystallized disputes over a traditional vision of women's role, etc. In communication campaigns, women are seen as water carriers and also, as the minority and vulnerable population. They will serve a large range of Communication for Development programs, non-profit association or private companies. Many exploited the figure women to bring awareness by transferring the poor women's condition into accessing to water resource in countries where tap water is common (See Box 6.2). They all want to show the urgency for action and the figure of women carrying at water has simplified their communication into one single visual representation.

Box 6.2 Women and Water at the chore of communication campaigns by UNICEF and charity: water

The UNICEF and charity: water created two powerful ads referring to the water chore and the relationships of women with water. These two ads are examples how the water chore can be at the centre of advocacy, discourses and creative stories to bring awareness on a vulnerable group of populations and on the universal challenge of water access.

In 2008, charity: water released a TV advert with the actress Jenifer Connelly. The actress starts a day by selecting large yellow containers. She leaves her house and walks the streets of New York. Several people are also carrying jerricanes and are all heading towards Central Park. The actress waits in line to get water and then get back to her house. She gives water to her children. The water is dirty and the children ask themselves if they should drink it. Jenifer Connelly smiles with content. The ad ends saying 1.2 billions of children drink dirty water. The emphasis is not put on women but the story plays with the regular scenery of women carrying water in countries without water access: some women are carrying water on their heads, some are using carriages, other eventually drop some water on their way back home… The water chore constituted the framing of the ad but the message focused on dirty water for kids. The goal is to convince donors to give money to provide clean water access to children.

The UNICEF released the TV commercial "Desperate" in 2009. Desperate is a reference to "Desperate housewife" as the character appears to be a woman staying at home and to "despair" and "desperate" as the real meaning of the story of the ad. The spot starts with a US mom in her kitchen when her baby starts crying. She reaches the fridge to fill the baby bottle with water. Since the water carafe is empty, she tries to open the water faucet. No water comes out and she rushes to the bathroom. The woman begins to look scared and takes her screaming baby into the street to get water from a public faucet that is closed with a lock. She walks down the street of the city looking around but nobody pays attention to her. She tries to get water from a water bottle in a public rubbish bin but the bottle is empty. She continues her walk and finally reaches a small water puddle in a corner of the road. She opens the baby bottle and fills it with dirty water. The baby drinks it and calms down. The video ends with a male voice stating that "thousands of women face this situation every day".

Both ads are found on Youtube under "Unicef Tap project Desperate" and "Jennifer Connelly in charity: water"

7 Global level: Framing stories

Studies on media coverage over water issues show that there is a need for more proactive communication from utilities, experts or authorities within the media (Dolnicar and Hurlimann, 2008). Despite the efforts of water professionals, stories about water tend to focus on topics that don't interest the public. Hoffman & Hoffman (2003) reported that articles published after the forum focused on 11.4% reported on World Water Actions, 11.9% reported on World Water Development Report, 11.9% on Financing Water for All (Camdessus Report), 0.5% on the Water and Climate Report.

Similarly, Tilt and Xian (2010) and Long (2001) concluded on the important focus of issues related to the government's perspectives Despite an important effort to reach the media with topics that interest water professionals. Recent studies on the framing of media on water also highlight the negative framing over water resources in particular, in relation to climate change. They tend to focus on "science," "catastrophe," "solutions" and "conflict"(Resource media, 2010). Others show the importance of covering disputes and conflicts (Miller *et al.*, 2009; Velayutham *et al.*, 2010), or science bringing the public to focus on risk (Corbett *et al.*, 2004; Driedger, 2007). Many professionals seek solutions and ways to interest the public. Non-profit associations that function thanks to donations also bring visibility to water challenges. They develop different initiatives in order to gain visibility working with the media. Media and non-profit organisations both question global framing of water stories. They underline the importance of framing stories rather than facts and to engage individual personally.

Both actors are building discourses that are impacting people on their understanding of water challenges. They hold a responsibility in disseminating messages and discourses that may be different to the reality of water resources management: media tend to reduce water to "water wars" and non-profit tend to promote one solution to water access through the promotion of wells. If they are addressing key issues, they might be reducing the vision and global communication over water stories.

7.1 MEDIA: A COMPLEX ROLE TO BRING AWARENESS

"Water journalism" is a new concept in advocacy, bringing in media as a key partner in the debate on issues of water scarcity, distribution, equity and ownership. Many networks participated in an initiative developed by the Water Sanitation Program: The Economist, Peru 21, El Comercio, Caretas, Semana Economica and Radio Programas del Peru (RPP). The UNESCO-IHE also conducted a wide program targeting the education of journalists in several countries. Many networks of "Water Journalists" are developing. They however, have difficult missions: to bring adequate information to the public and to recruit new journalists to their cause, etc. Kajsa Dalrymple shows the role of opinion leaders in promoting environmental behaviours in relations with mass media and campaigning. Fredrick Mugira summarizes the difficulties and missions of African Water Journalists. Lis Stedman explains the challenges of writing water stories interfacing technological aspects with easily readable content to the public.

7.1.1 Mass Media & Opinion Leaders: Promoting Environmental Behaviours

By Kajsa E. Dalrymple
Assistant Professor in the School of Journalism and Mass Communication,
The University of Iowa, USA

Concern for our environment has been a topic that has received increased attention over the last five decades. Although discussions of issues such as pollution, DDT, overpopulation, and energy have piqued public interest in environmental issues over the years, recently discussions regarding the concerns of global climate change have helped turn public attention to environmental issues. Now, topics such as alternative energies, melting ice caps, and erratic weather are commonly discussed in public forums and mass media messages in relation to global climate change and its effects. Similarly, as these discussions regarding rising temperatures, natural disasters, and greenhouse gasses have developed over the last 50 years, it has allowed less salient issues to come to public attention. For instance, groups promoting concern for water quality, the future of our water resources, and sustainable consumer behaviour now have greater opportunities to share information and voice concerns to the public, and recent polls show that water issues are now topping the list of public environmental concerns (Saad, 2011).

At the same time, however, research indicates that the American public struggles to make sense of scientific issues, including those related to the environment (Nisbet & Myers, 2007). Similarly, survey research has indicated that only a small portion of Americans have adopted personal behaviours that may aid in reducing the threats to the environment (Maibach *et al.,* 2008). As environmentalists continue to dispute the most effective methods for encouraging public awareness of these issues and, subsequently, encourage public adoption of preventative behaviours, it is necessary to reflect upon two broad questions: First, what sources do the public turn to for information regarding the environment? And, second, what types of communication are most influential in encouraging preventative environmental behaviours at the public level?

In order to address these questions, it is necessary to consider the relationships between media use, opinion leaders, and environmental attitudes and behaviours, in order to tease out the potential flow of influence from mass media (at the macro-level), opinion leaders (at the meso-level), and individual behaviour (at the micro-level).

The idea of multi-level communication effects has been discussed throughout a broad range of communication, political science, and sociological and psychological literature. Much of this literature has focused on the potential relationships between mass media and the individual. Specifically, scholars have pointed to two types of within-level relationships – macro-macro (e.g., relationships within the media industry and effects on social characteristics), and micro-micro (e.g., relationships between individuals) – and two levels of cross-level relationships – micro-macro (e.g., relationships between individual behaviour and aggregate social effects), and macro-micro (e.g., relationships between media and the individual) (see, Coleman, 1986; 1987; 1990; Pan & Shah, 1991).

More recent research has attempted to apply a multi-level approach to the potential relationships between media and the individual, however, research still struggles in teasing out the true nature of these cross-level effects. For instance, research examining the concept of social capital often focuses on the micro level with individuals used as the unit of analysis, rather than exploring the true multilevel nature of the concept (e.g., Putnam, 2000; Shah, 1998; Uslaner, 1998). Similarly, research investigating the effects of interpersonal discussion on social participation has only recently begun to consider potential causal relationships between individual characteristics and aggregate behaviours (Klofstad, 2007). Considering this, it is necessary to consider the potential cross level relationships that may exist in the processes related to environmental behaviour change.

Within the context of communication and behaviour change, the role of effective communication becomes quite clear. Communication, both mass and interpersonal, plays a key role in the improvement of public perceptions and attitudes toward environmental problems (Stamm *et al.,* 2000). Similar to research examining the concept of social capital, the processes behind behaviour change appear to take place across multiple levels of analysis. Specifically, research suggests that the potential for influence appears to take place across three levels of analysis: 1) at the micro, or individual, level, 2) at the meso, or group, level, and 3) at the macro, or mass, level. In this case, the macro communication level is represented by the dissemination of information by mass media; the meso network level is represented through individuals' interactions with opinion leaders; and the micro individual level reflects citizens' levels of motivations, attitudes, and knowledge to facilitate their behaviour choices.

At the broadest level, research has pointed to the large role that mass media place in influencing public opinion and behaviours. The very nature of mass media is to reach the many, or the large aggregates of consumers. As a result, the relationships between mass media and the individual are often one-sided and impersonal (McQuail, 2005). At the same time, however, audiences often look to mass media for portrayals of cultural phenomena, societal norms, and information about current events (Altheide & Snow, 1991; Carey, 1975; Westley & MacLean, 1957). Additionally, research points to the important role that mass media play in the formation of social integration and identity. For instance, Putnam (2000) has shown the potential impact that television use can have on individual levels of participation and perceptions of "social capital." Similarly, research has indicated that mass media can bring individuals together by providing the common values, ideas and information necessary to form social identities (e.g., Stamm, 1985; Rogers, 1993). Much of this research focuses on theories related to the content and structure of mass media, such as agenda setting, framing, or cultivation (Bryant & Cummins, 2007). However, when considering the impact mass media may have on public behaviour, it is also necessary to consider the influence of those at the group and individual levels.

At the meso-level, research points to the importance that social groups can have on individual actions and opinion. Much of this research focuses on the important role that social networks may play in the formation of societal perceptions. In particular, scholars investigating social networks argue that the ties within groups not only facilitate the flow of information within a group but can also play an important role in establishing social norms between groups of people. Within these groups, opinion leaders are often the greatest contributors to the interpretation and integration of information (Bandura, 2001; Brown & Reingen, 1987; Granovetter, 1973). As a result, these groups of opinion leaders tend to be very influential in getting people to change their attitudes and behaviours because they share similar traits to those they influence. Both health and environmental research supports these arguments with consistent findings pointing to the importance of opinion leader groups in encouraging positive behaviour change or compliance. For instance, a recent study showed that the reputational benefits of purchasing "green" environmentally conscious products outweighed the fact that such products are often of lower quality, convenience and performance (Griskevicius, Tybur, and Van den Bergh, 2010). In conjunction with the effects of mass media use, it is clear that factors at both the mass and group level can be very influential in shaping public perceptions and behaviour.

At the same time, however, it is also necessary to consider factors at the micro-level. Similar to the effects that social networks of people or opinion leader groups may have on behaviour, individuals are also very influential in shaping public behaviour. At the simplest level, individual relationships are based on a set of expectations that two people have for their behaviour based on the pattern of interaction between them (Knapp & Miller, 1994, for an overview of interpersonal communication). The nature of a relationship, therefore, is defined by the communication between its members. As individuals communicate, they are continually creating, reinforcing, or changing the expectations within that relationship (Watzlawick, Beavin & Jackson, 1967). As a result, individuals often try to reduce uncertainty or anxiety within their relationships by self-monitoring their actions and modifying behaviours to suit others (Gudykunst, 1995). Consequently, our individual relationships are often very influential over our perceptions of cultural norms, appropriate behaviour, and societal expectations.

In contemporary society where scientific progress is rapid, media are a pervasive source of information about science and the environment, and research has continuously demonstrated the important role of media use on public understanding of scientific issues (Wagner *et al.*, 2002; Nisbet & Scheufele, 2007; Nisbet & Goidel, 2007). Research in the field of communication has shed light upon several explanations regarding why the media are so effective in shaping environmental attitudes. Some studies have established that there are large gaps in media coverage of environmental issues, claiming that media focus more on dramatic events related to social and economic impacts rather than systemic concerns, thereby heavily influencing public perceptions of environmental risks (Friedman, 1990; Hansen, 1999; Ostman & Parker, 1987; Holbert *et al.*, 2003). Following this argument, agenda-setting research has also claimed that media coverage is responsible for directing public attention both towards and away from environmental issues, due to trends in episodic and sensational news coverage (Atwater *et al.*, 1985; Ader, 1995; Holbert *et al.*, 2003). Although the effects of mass media on environmental attitudes and behaviour may be mixed, a major assumption underlies much of the research regarding environmental media consumption – mass media not only encourage the public understanding desired by environmentalists and policy makers, but can also influence subsequent public action.

Based on our current understanding of human behaviour, we know that a behaviour is guided by existing beliefs about aspects of the behaviours such as its consequences, the expectations of others, and individual control over performing the behaviour (Ajzen, 1988; 1991; 2002). For instance, a behaviour like converting your home to more environmentally conscious standards of living, would be guided by beliefs about whether adopting those standards is an effective means of reducing use of non-renewable energies, whether you believe that others think you should adopt those standards, and whether that adoption is an affordable and otherwise doable behaviour for you to perform.

According to the theory of planned behaviour, personal values and beliefs are likely to vary across different audiences (Fishbein & Ajzen, 1975). As a result, strategic messages that target specific group and individual existing beliefs (such as the outcomes related to a specific preventative environmental behaviour) may be effective in producing behaviour change. Therefore, it is necessary to consider the importance of media as a tool for informing the public, and the role of mediated messages in the context of these broader behavioural factors.

Additionally, it is necessary to consider the role of social influence in the process of environmental behaviour change (Dalrymple, Shaw, & Brossard, forthcoming). We often rely on our perceptions of others as a way to determine whether we will engage in a particular behaviour. This trend has been highlighted in much health and environmental research highlighting the importance of subjective norms in encouraging behaviours such as maintaining a healthier diet (Lytle *et al.*, 2003), using condoms in order to prevent sexually transmitted diseases (Fishbein *et al.*, 2001), and recycling (Terry *et al.*, 1999). Despite the fact that mass media are widely recognized as extremely important sources for scientific information (Nisbet & Goidel, 2007) and as powerful instruments in shaping our attitudes toward many social issues (Allum, Sturgis *et al.*, 2005; Brossard & Shanahan, 2003; Priest, 2001), interpersonal communication is also an important factor in shaping public behaviours. Additionally, it is the interaction with opinion leaders within one's social network that is often most influential in behavioural decisions.

As previously mentioned, scholars have long recognized the importance of opinion leaders in shaping social norms, informing fellow citizens, and altering behaviour (Asch, 1956; Cialdini, 2001; Crutchfield, 1955; Lakin & Chartrand, 2003; Mackie, 1987). This research has shown that there are individuals in society who become more involved in issues than others (Keller & Berry, 2003). As a result, these individuals tend to fill the important role of passing along information to their social networks and upholding social norms. Subsequently, these individuals tend to be more persuasive in convincing others whether or not to adopt certain opinions and behaviours (Katz & Lazarsfeld, 1955; Keller & Berry, 2003). While these may not be recognized as individuals of power within their communities, they still can serve as champions of change. Additionally, this influence may occur unknowingly when these opinion leaders are conducting daily behaviours such as offering advice or recommendations, participating in local forums or discussions, or engaging in debates about various opinions and behaviours. As a result, these influencers can aid in the encouragement of environmental behaviours while remaining completely unaware of their influence on others (Nisbet & Kotcher, 2009; Weimann, 1994).

Taken together this research can tell us quite a lot about the dynamic nature of our contemporary media and public perceptions of environmental issues that face the nation. The most noticeable conclusion that can be drawn from this summary is that mass media are not only necessary to the process of behaviour change, but that the effects of media use are varied across social groups. Although the general public may regard mass media as a reliable source for information about the environment, opinion leaders may be more critical source of such messages.

Practically, we can use this information to inform both large-scale campaigns and grassroots efforts to encourage behaviour change. While these results may direct campaign strategists to rely on integrated marketing strategies, they also offer insight as to how opinion leaders can be used to mobilize the public. Campaigns that intend to utilize opinion-leaders should be wary of not only message construction, but also the current social dynamics surrounding the topic of interest. Empowering opinion leaders as champions for change, in addition to involving them in the process of message construction and communication strategy will be critical to the success of any future communication efforts regarding the sustainability of our natural resources and global environment.

7.1.2 Making the Unfamiliar Familiar – a Narrative by African Journalists who are Enhancing Water Awareness Through Social Media

By Fredrick Mugira[1]
Journalist, Uganda

Reporting on environment and water in particular has not been a favourite subject for most African journalists.

Water stories remain inadequate in the African media even though coverage of these issues on the continent has increased over the last decade. There are few stories of this nature in Africa media that are again buried deep inside most newspapers. It is not common for water stories to attract the public's attention the way political and sports stories do unless such stories are controversial or negative.

Why Journalists Don't Cover the Water Story

Journalists, media managers and media owners claim, reporting on water issues and environment in general requires a lot of money which some private media companies on the continent lack. This is compounded by the political and sports interests that shape the media coverage on the continent. Other reasons further to this are that such issues are often complex and boring.

True. Water issues are multi-faceted. They actually stretch to issues of science, business, economics, culture, and health amongst several others. Therefore, good water reporting requires flexible minds that can readily understand these issues, how they interweave and how to simplify them for a common audience.

In many cases water reporting in Africa is mainly superficial interpretation of water issues and the stories are full of jargon. I attribute this to a silent refusal of journalists and their editors to style up when it comes to reporting on these issues.

---

[1] See Deanne (2005) and Karembu & Kiai (1999).

Corruption in some African countries is breeding a lack of transparency in the water sector. In the end, officials in the sector refuse to share information with journalists because they fear acting in open manner. They conceal information and refer journalists here and there causing some journalists to find difficulty in verifying information, relying on a single source and in the extreme cases giving up such stories. Relatedly, some corruption officials in the water sector give 'brown envelopes' to journalists so that they keep off stories and do not hold them accountable for their wrong actions.

The political interests that shape the media coverage in Africa further compound those problems. In some African countries, the media are still strongly influenced by governments. Journalists are compelled to serve the, often wrong, interests of governments in power. Why? Because the role of journalists should go beyond what governments, NGOs or individuals like or dislike. Journalists should remain responsive to the interests of their audience.

Critically related to this is the fact that some media on the continent are increasingly focusing on manipulating opinion, shaping public discourse, and promoting the consumption of products. This is therefore shaping the content of media on the continent to focus mostly on the sections of society that have disposable incomes and leaving out stories of rural, poor and otherwise marginalized people on water.

Inadequacy in covering water stories in Africa could also be related to lack of exposure to concepts in water systems among journalists. Related to this, water journalism is often not seen as a priority that influences the nature and amount of media coverage given to water issues. Perhaps journalists should take some degree of specialization in water issues.

Journalists Making the Unfamiliar Familiar

There is still hope. There are some journalists on the African continent who have passion for reporting on water and they believe nothing can stop them. They are taking advantage of the free social media to form networks and platforms such as blogs for enhancing water and sanitation awareness.

One of such networks is Water Journalist Africa (WJA) - a network of journalists in Africa who report on water and sanitation. It started on 23rd March 2011 in Cape Town South Africa by a group of African Journalists who report on water under the guidance of UN-Water Decade Programme on Advocacy and Communication. This was during the 2011 UN Water Day.

The network started with less than 15 journalists and widened to comprise 100 journalists from various countries on the continent, before its first birthday. By the end of April 2013, the group had over 200 active members. These journalists are using their blog www.waterjournalistsafrica.wordpress.com to facilitate communication between the public, policy-makers and technocrats on the continent. Membership and registration is free.

The Water Journalists as they call themselves have been able to start an e-paper they named WaterSan Perspective. This e-paper contains water and sanitation stories, it is helping them to promote water awareness, enhancing interaction and improving their ability to report on water and sanitation.

Members of this network believe that these stories do not end in the online community but have an impact on the offline world in the African villages through information diffusion.

By the end of April 2013, approximately two years after the establishment of this network and it e-paper, the WaterSan Perspective had 57,131 all-time views, 1,316 views on its busiest day and 179 comments.

The network is helping to empower journalists through various trainings. For example, in July 2012, Water Integrity Network and Transparency International Kenya trained all members of this network from Uganda, Kenya and Tanzania on Water Integrity in Nanyuki, Kenya.

But what is outstanding is that through the network's social media platforms such as Facebook and email groups, journalists who belong to this network have been able to create friends, network and share knowledge with them for the betterment of their profession.

The network is also exposing members to new and emerging concepts in WASH journalism and reporting on water in particular. Subsequently this is aiding them to report on water issue with ease.

This network was recently chosen to represent African journalists on the African Development Bank's Regional Coordination Committee (RCC) for the Rural Water Supply and Sanitation Initiative (RWSSI). This committee was launched in Tunis, the capital of Tunisia in March 2013.

With the guidance of the UN-Water Decade Programme on Advocacy and Communication, the Water Supply and Sanitation Collaborative Council (WSSCC), UNESCO-IHE and the African Development Bank, the network is growing day by day.

7.1.3 Writing About Water: The Changing Face of Communication

By Lis Stedman
Journalist, UK

Water: a diverse sector

The greatest challenges in writing about the water sector relate to its sheer size and complexity – it is possible to move directly from writing about a highly-technical subject such as reverse osmosis to a discussion of water company regulatory capital values, and then on seamlessly to a topic such as water, sanitation and hygiene (WASH) interventions in Africa.

It is only necessary to look at the number of International Water Association (IWA) Specialist Groups to realise how very many topic areas this sector presents. In framing the challenges that accompany each of these it is necessary firstly to remember that while one is writing for an expert audience, the audience as a whole may not have specialist expertise in the particular subject that is the focus of any one article.

This means that any writer in this field faces a significant challenge themselves – communicating sometimes highly-complex ideas with sufficient clarity and depth of detail for an audience, some of whom have a high level of expertise in the subject and others who, while intelligent and deeply immersed in other aspects of the sector, do not.

The balancing act, and how to achieve it

In achieving this delicate balancing act, a writer must be able to elicit sufficient information and possess enough general background knowledge to produce an article that is enjoyable for this varied and knowledgeable group. This is not easy. Even to know which questions to ask an interviewee with a vast amount of expert knowledge requires a solid understanding of the issues and challenges that affect that part of the sector.

Writing "outwards" to a non-expert, non-technical general public audience is even more daunting because not only is it necessary to thoroughly understand the sector and its wide array of issues, it is also vital to be able to step back from this knowledge to assess what the "man on the Clapham omnibus" knows of and feels about these matters.

The water sector can be the subject of intense emotion – one need only look at the huge and sometimes overwhelming challenges that private companies have faced in providing water and sanitation services around the world to understand that that is true. It is such a vital sector that its activities affect people in a very direct and personal way, and this significant level of concern must be understood and respected in order to communicate well.

The changing language of water and communication

In terms of how writing about the sector has evolved over the past 30 years, there have been many changes – for instance, huge advances in technologies such as membrane bioreactors, UV and reverse osmosis (RO) systems that have seen these move from being niche, expensive options to widespread adoption.

Climate change has also added an unwelcome twist, and an increasingly urgent debate about water scarcity has developed since its effects have become apparent. One of the most pressing needs is to communicate better with businesses the many advantages of reducing their water use, given this driver, and (on the other side of the coin) to disseminate messages related to flooding, for instance by highlighting the debate about building on flood plains and the benefits of Sustainable Urban Development Systems (SUDS), known as Best Management Practices (BMPs) in the US.

The language of the debate has also changed – concepts such as water, sanitation and health, disaster resilience and the water-food-energy nexus are proof of a growing awareness of how impacts from one sector, event or action affect others. This relatively new realization of the inter-connectedness of all things provides commentators with exciting opportunities moving forward.

The net result is that water is no longer viewed as an isolated, specialist sphere – its fundamental relationship with many other life-critical issues is now laid bare, and those who are writing about the sector are among the early explorers charting developments in this new and exciting domain.

Ways forward for writing

These developments, I feel, make the task of raising public interest towards water somewhat easier. Generally one senses from the growing number of articles that appear in the general press that there is increased concern and awareness of water-related issues. Although these can sometimes be negative and ill-informed, they provide a catalyst for discussion that hopefully, will enable more knowledgeable comment to thrive.

In terms of recommendations for helping the public to engage with water issues, I would suggest leveraging this increasing awareness and level of concern. There is a need for articles that present these issues in an easily digestible format, and which empower people to join the debate by providing the unbiased, impartial and endlessly interesting truth.

7.2 NON-PROFIT: SIMPLE MESSAGE, SIMILAR DISCOURSES

7.2.1 A Simple Message: Water Kills

With Renaud Douci,
Solidarités International, France

When we decided to share the successes of our activity, it was really something new for us. We are usually doing humanitarian work and we are more used to working in the field. We are more like engineers or planners rather than communication strategists! At first people in our association didn't see the value of communicating to the public. We had to simplify our message and to find ways to engage the public. It couldn't be about our activities or technical skills; people don't really understand how we work. We also had to find something original and different from other French non-profit associations. We decided to focus on water since more than 70% of our activities are dedicated to water. We were the only French association involved in water projects. We also wanted to raise public awareness on the lack of water access and the number of death due to lack of access to clean water. It was a simple message but we needed to bring this message to institutions to have more funding opportunities in order to develop our programs. We also observed the lack of awareness from the media and the public about the first cause of death worldwide : the lack of water access.

It was difficult to make our choice to address such an issue. In 2006, we first conducted a campaign towards media for the World Water Day. We wanted to create a sort of "buzz", to gain the media's attention. We staged a small event which featured actors "dying" in front of a fountain. To accompany the event, we distributed bottles filled with dirty water labelled with different diseases: cholera, typhoid, etc. It had an impact and we established our initial relationship with the press. It was a good experience and our message was very simple: water kills!

After this successful experience, we decided to continue organising campaigns each World Water Day. We made several campaigns and we launched our petition in 2007. Our strategy was to encourage people to sign the petition and bring it to policy makers. Our petition included facts and several commitments to improve water access such as the recognition that dirty water is the first cause of death in the world, that the UN reviews their data about water access in the world, and to increase funds for water access. In 2008, we gained 50,000 signatures, etc. It was a good result but we had to gain more visibility.

In 2010, we really started to gain a real visibility, etc. We set up a "water wall" in front of one of the Louvre entrances in Paris (Figure 7.1). It was a well-located site, in the centre of the city and where many people visit. Water was falling with written phrases such as "I kill more than cancer, AIDS, or guns, etc. I am WATER". It really helped us to have more visibility in France, etc.

Figure 7.1 Water Wall in Paris, 5 meters high and 6 meters long (*Source*: Solidarités International).

The year after, we launched the campaign "Water Ink", our first world success! The spot is a simple idea: to use the dark colour of the ink with the liquidity of water (See Figure 7.2). After, we needed to develop a strong image. We choose the face of a child that would embody the idea of danger towards those most vulnerable. We worked with an artist to shoot the spot. He took one year to make the movie! He was very dedicated and wanted to bring a powerful message with vivid images. And the result was quite incredible. The spot worked well in France but actually, today; the spot is more famous worldwide! We have more people from the USA, China and the UK watching it on the Internet. We also won several prizes and it gave us more legitimacy in bringing important messages to policy makers.

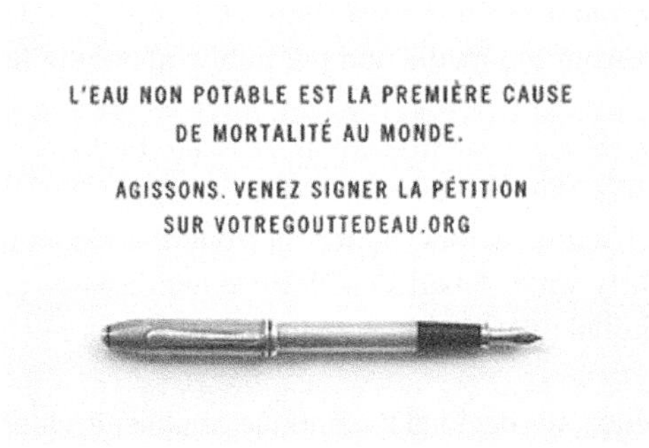

Figure 7.2 The ad "Water Ink" combines the ink, the pen and the petition (*Source*: Solidarités International).

In 2012, we lobbied to be present at the World Water Forum. We were the only humanitarian association attending the official ministerial session and we were able to deliver our messages. We reached 100,000 signatures thanks to our campaign on social media, the "Minute campaign" (Figure 7.3).

Figure 7.3 The "Minute", 7 people die from dirty water with as dedicated website, www.laminute.solidarites.fr (*Source*: Solidarités International).

For this spot, the concept was to engage people thanks to social media. We developed three TV spots with the same approach. We had three people waiting, one was waiting while phoning to a customer services line, one waiting for her nails to dry and another one drinking a coffee. Then, we showed that during this minute, 7 people died due to dirty water. At the end, people were invited to sign online. It was also very successful; we had 35,000 signatures in less than 3 months. It was pretty difficult to have people sign, but we reached 100,000 signatures only a few days before the start of the forum.

For the opening session of the World Water Forum, we were able to bring the petition to the officials of the Forum. With the petition, we communicated on our estimations of water access. We actually gained visibility after the UN published their data on water access. We contradicted the data saying they underestimated our petition and that the media used our data. Actually, I didn't understand the strategy of the UN. We still need more funding and more engagement from all stakeholders to bring water access at the local scale.

I believe that communication today in the water sector is very institutional. We put too much focus on the right to water from a legal perspective. There is also a big focus on MDGs, etc. People don't understand what the MDGs are but they understand that we need to stop the number of diseases and deaths due to lack of water and sanitation access. We need to bring messages from a human perspective not from the science side or from legal aspects. Look at the successful campaigns today. The non-profit association Action against Hunger is very successful because they use simple messages. I also really envy the communication of the charity: water. They have real skill in bringing simple message into images. This is actually what I would need as a professional. I would like to have recommendations on how to communicate about our local activities. Do we provide infrastructures or help communities in urgent situations such as wars or a natural catastrophe? How can one communicate about those programs without too much complexity? How can we speak to people at their level with simple and powerful images? It is all the more difficult that we need to bring a new creative campaign each year!

7.2.2 Similar Discourses: Overview of Non-Profit Water Campaigns

Non-profit associations rely the public to get funding through donations and to gain financial support from larger institutions and funding agencies. Non-profit associations usually want to bring a political message and advocate for one major cause: water and sanitation access. Several characteristics can be found in the discourses of non-profit associations:

- The figure of the leader
- The vulnerable population
- The facts
- The engagement/the cause
- The world solidarity for this cause
- The way to address the challenge

In many non-profit associations, the story of the founder is a key element to the building of communication campaign. Charity: water has launched it whole strategy based on the story of their founder and how he decided to give away his birthday to gain money and to help people in Africa gain access to water. Scott Harrison is also very present in videos at the local scale; on the different communication materials (social media, websites, letters, etc.) the founder is himself, a celebrity known for his engagement. Matt Damon for water.org, Edwin Broni-Mensah for Give me Tap, Sihamedi Idriss for Baraka City, etc. They are at the centre of the campaign to show the serious of their engagement, to embody the cause and to make people adhere to their cause and engagement.

Their personal story to provide water access to people plays an important role in promoting the activities of their non-profit association. The embodiment of the individual engagement that bring global change will convince the public that their individual donation will also act as an engagement and a solution to water access

Along with the story of the founder, the story of vulnerable people is often used. Usually, the story of the founder began with an event with the vulnerable targeted public. They met some women carrying water or saw children dying from dirty water and became aware of the issue. This striking personal experience made them realize they had to do something for them. The vulnerable population becomes the centre of the action. The description of their situation creates empathy and most video or tools will show their poverty but also, their capacities to adapt and to cope with the situation. The depicting of the population remains positive, optimistic and full of hope in order to let possibilities for changes and actions.

As seen in Chapter 3 about water visual identity, visual data will help the discourse: maps, representations of men and women, numbers or drawings of water infrastructures, etc. Non-profits are widely using such tools to turn their actions into concrete processes and programs. They show the chain of value and how each individual can make a contribution. The major objective is to show how one contribution – one small amount of money – helps to solve the problem that is presented. On the internet, websites will also be designed to present the activities step by step. The visitor will scroll down through pages that will show how the non profit addresses a global challenge (dirty water, daily diseases); works at a daily basis to solve the problem and how many people have already joined to participate to the solution. This strandardized discourse from non-profits mainstreams a global framing of water as a cause of disease, it also conveys one type of solution, the construction of wells. Such global discourses are contradictory to tap water discourses. Despite the clear difference of situation, the depicting of water as a cause of diseases will have an impact on donors who are mostly living in countries with safe tap water access. As such, the solution remains the building of wells to differentiate between tap water at home provided by water pipes and infrastructures, and with wells in countries without water access to solve the challenge of dirty water. It however questions impacts on the public's perceptions of tap water, and the impacts on the sustainability of financing water access through wells and fountains rather than water infrastructures.

8 National-Regional Level: Interfacing Communities

The water sector is known to be fragmented and wide. As stated, many stakeholders are involved in water cycle management. Those numerous communities need and seek mediation and interfacing. Each community holds their own priority of usages and agenda. This chapter focuses on public actors at different levels. The objective is to show how different groups need to adapt their language and methodologies to reach better water management and knowledge brokerage for the public good and protection of water. Two different approaches are scanned: the internal process among the water sector e.g. a version of internal communication and how professionals within the water sector need to better communicate, mediate and collaborate; the external process with a main target, the general public or individuals. For each group, a specialist brought his/her experience and recommendations.

8.1 INTERFACING COMMUNITIES WITHIN THE WATER SECTOR, INTERNAL COMMUNICATION

Philippe Quevauvillier shows the importance of the Science-Policy interfacing from his experience at the European Commission. Beatriz Medina shares her experience on an EU funded project that aims at interfacing research with the market. Jean-Philippe Donzier brings his experience for INBO and the importance of interfacing stakeholders at the Basin level.

8.1.1 Science-Policy Interfacing. The Experience Of EU Funded Projects in Communicating Water Research and EU Water Policies

By Dr Philippe Quevauvillier,
European Commission, Belgium

There has been debate and discussion over most of the last decade over the disconnections between the research and the policy making process. The European Commission launched its first workshop on science and policy interfaces in 1998 in order to discuss how to better communicate research results towards policy makers and vice-versa and how to communicate policy and EU directives towards researchers and practitioners. We rapidly pointed out the gap between science, research and policy makers. This gap was identified as a clear obstacle to the implementation of major EU policies including the Water Framework Directive. It also represents a barrier to more innovation and applications of research results into the market.

After 10 years observing the different communication strategies set out by the different EU projects, I truly believe what is needed, is a kind of a pyramid of communications, reaching policy makers at the top of the pyramid and down to the practitioners at the level where the research findings will ultimately be applied as perhaps new technologies, methods or approaches.

We firstly need to ensure that practitioners agree that the research is technically relevant and they would be potentially interested to apply new methods and new technologies, for example for their policy implementation tasks. This bottom-up approach is, however, not sufficient as many (convinced) practitioners might not have the possibility to apply innovative approaches without the agreement of "higher" decision-making levels. Therefore, we need also to go to the top to tell the people who are deciding about the use of new technologies that it is relevant for them to consider, which requires a different type of communication that is more oriented towards the policy relevance.

From the above, it appears that this communication needs the research findings to be distilled into appropriate formats for the respective audiences, requiring people specifically charged with this task. I think it calls for "mediators" (translators of science), i.e. specific people who will act between communities, and ensure that the right language is being used at various levels. I consider that this mediating activity between science and policy is a function that needs to be strengthened at all levels – EU, transnational (in the case of shared river basins), national (member states), and regional levels. There are timing implications too. Such as ensuring that when research starts in a given area, first of all we need to establish a science-policy dialogue at a very early stage of the project. In the past, I have actually been in a situation where I was a policy maker receiving reports or scientific papers at the end of a project, which was absolutely useless because it was too late. We should work on the basis of this pyramid of communications with a "cascade effect" to establish mechanisms to set the best methods and to communicate and transfer the research output to the users. A platform, which allows such mechanism, is the WFD Common Implementation Strategy, which includes consultation and coordination at different levels (from the top, EU Water Directors, to a coordinating group down to expert groups). It will be beneficial especially at EU level where research is fragmented and researchers work in competition, i.e. joint communication efforts will enhance the visibility of research and the critical mass but this needs to be done by "independent actors" (mediators). Fragmentation indeed limits the research efficiency. For instance, if you take a subject area such as flood research, just three or four projects may involve something like 80 institutes across Europe. It constitutes a very big critical mass that is possibly not used properly owing to a lack of synergies (a recent science-policy event at the FloodRisk2012 conference has clearly pinpointed this drawback). Policy makers, citizens, research institutes would benefit from having an overview of the state of the art of research that the scientific community could provide while still being acknowledged for their respective research findings.

Certainly the European Commission is in a position to influence what happens i.e. requirements are introduced into research project contracts setting out communications requirements. This could be improved by communicating research outputs in a "policy-formatted" and systematic way (progress has been done recently with the production of "science-policy" factsheets). Member states could then ensure that findings are translated into the national languages and then relayed to practitioners. This is an important aspect since projects may include pilot or demonstration work at the sub-national level where respective member states may not have been directly involved. The ultimate step could be to have technical guides describing how to implement new findings, which would match policy guidance documents requirements such as those developed by the Common Implementation Strategy.

We need to turn these ideas and discussions into practical measures without necessarily passing by legally binding instruments to implement this process. The European Commission may facilitate, provide recommendations and further requirements but a real change towards implementation of science findings at the most appropriate level is also a question of willingness and change of working habits. At EU level, researcher and policy-makers are now progressing towards more interfacing through a "Science and Policy Interfacing" initiative as part of the Common Implementation Strategy. This activity of interfacing facilitates the communication between research, policy-makers and practitioners. The need has been recognised, the concretisation is now key.

The major challenge is now to ensure that this type of interfacing will become current working practices at EU, national and regional levels. This requires a change of mentalities so that it becomes natural to different types of target groups to work together and to understand the specificities of communication requirements, processes and disseminations. I truly believe it will support a stronger innovation among EU projects and will bring further added value to research. Through a dedicated Water Communication processes, we are changing of paradigm. Ultimately, it will ease the process of applying water-related research into the market but also to raise awareness of the general public and change behaviours towards better protection of water resources. It brings more understanding of what communicating water means in particular, to differentiate targets, messages and interfacing phases.

8.1.2 Research - Market, Can We Speed-Up The Knowledge Transfer Process?

By Beatriz Medina,
Amphos 21 and Water RtoM partners[1]

The implementation of the Water Framework Directive (WFD) and its daughter directives, as well as the Floods Directive, needs new knowledge and know-how. This concern is shared by many stakeholders, who would like to boost the transfer of research outputs to the real life, even if most of current initiatives are focused on the policy-making step, not the day-to-day policy-implementation one.

When looking at the impact of research on water management practices, it is demonstrated that the connection between research and the policy process is not efficient ; as stated by FUNDETEC, a FP6-project, final report in Dec.2007, "the typical length of time needed to complete the development cycle (in the water sector) is 10 years"; it means that research commissioned today will impact water management practices within about 12 years, far after the milestones of the Water Framework Directive (2015, 2021).

The Water Research to Market project addressed this challenge to speed-up the transfer of research outputs to practitioners. This four partner's initiative (International Office for Water, France; Amphos 21, Spain; Gdansk Water Foundation, Poland; and Fundatia Centrul de Formare si perfectionare Profesionala in Domeniul Apei ,Romania) has been running for 3 years, started up in September 2010 and ended in September 2013. Results were gathered from the analysis of more than 200 research outputs from both national research programmes of the four participating countries and EU level research and innovation projects. The analysis was framed under the perspective of innovative potential of the water related research output and under the experience of different dissemination tactics adapted to each individual target audience (topicial seminars, workshops, exhibitions, e-seminars, etc.). Evidence from all those activities could be integrated into a guideline document, which aims to pose changes on the key stakeholders' attitudes in the innovation process. The main focus was fostering a proactive and effective dialogue among all the stakeholders among their existing transfer scheme.

Water RtoM carried out as a service linking authors' outputs (named researchers or knowledge makers) and users' outputs (practitioners or knowledge users). Therefore the addressed target groups were the practitioners whose most of them come from basin organizations of sub-basin authorities, water users, and suppliers of technologies and also researchers, knowledge transfer institutions and research funding bodies.

The main challenges faced by this initiative have been on how to motivate exchange of existing new knowledge from the research side in a two-way direction with the potential users. Determinant factors such as the willingness of stakeholders to engage in this exchange process or for them to change their language in a common approach have been raised. Further key challenges were with regards how to analyse a research output coming from multi- dimensional and multi-disciplinary databases in a marketing oriented vision. The amount of information from research faces the efficiency of knowledge brokers in finding key information useful for potential users.

[1] LIFE09 ENV/FR/000593

Success stories within the project (closing the innovation cycle) raised the importance of having a knowledge transfer boost for a better mutual understanding process to complete transfer of knowledge into the market.

The partnership activity coming from the engagement work with a Liaison Committee (a board of experts within the project) highlighted the importance of "working together" to assess if a research output is ready to be taken up, to better find the ways to promote it, to identify and reach potential users, and to identify and update market needs. This partnership can take place at European level but is essentially needed at regional and national level.

The new paradigm on how to better transfer knowledge in the water Science-Policy-Industry gap should be addressed in accordance with the changes in the needs of the active social system of the stakeholders, adapting transfer strategies and fostering dialogue, thus leading to the "speed up of the transfer of the research results" (Figure 8.1).

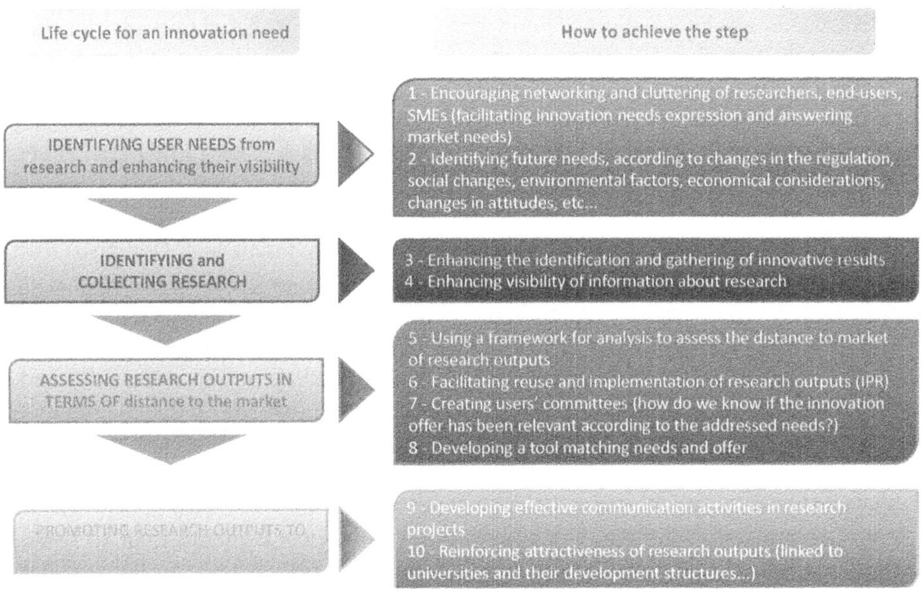

Figure 8.1 10 Key Recommendations from Water RtoM.

Tools, best practices and recommendations are based on the "output life cycle": Identifying the needs of the water sector in research → Identifying and collecting research outputs → Assessing research outputs in terms of their distance to the market → Promoting research outputs to the users for the re-use market uptake (Water RtoM, 2013).

8.1.3 Basin River, An Approach for Interfacing Stakeholders

Interview with
Jean-Philippe Donzier,
President of the INBO, France

INBO is a large network of water stakeholders created in 1994. Our website counted more than 1.3 million views last year. People are mostly looking for information about events organised on water such as conferences, training sessions, seminars, etc. We target institutional actors and professionals from the water sector; we don't aim at reaching the public. We don't put much emphasis on marketing or design but we take care of providing useful information in a rapid manner. Our approach is to give access to materials about past events, such as conference proceedings; and to allow exchanges on successful programs and best practices. We want to facilitate the management of water resources from a river basin perspective.

When we started in 1994, the approach was not well known, but today, the Basin Level approach is known and approved. It has been widespread throughout the world and now, the basin level approach is implemented in 80 countries. We saw a lot of progress in Africa where most of the great African rivers have implemented Basin authorities to facilitate cooperation between States sharing a river. The example of the Senegal River is particularly interesting.

The Senegal River, a successful story

Four countries share the water of the Senegal River. They rely on the Basin Authority (http://www.portail-omvs.org/) to manage a call for water infrastructures. They have programs to manage projects related to energy in order to guarantee the provision of sufficient energy resources to the four countries; to rule transportation and navigation; and water supply and distribution for agricultural purposes. Irrigation is very important for the neighboring countries that lack water resources, as is the efficient management of water resources. The four countries understand the added value of collaborating together.

In terms of communication, they provide the overview of the basin thanks to maps integrating several data that provide an overview of the Senegal River. All those data facilitate the management of water resources. They are collaborating with several networks and they are also very present at the international level. They work with several funding agencies worldwide. Their success story has inspired other countries in the region in particular, the Niger River. Five years ago, the seven established the Niger Basin authority that now, has its organization. They engaged a process of consulting stakeholders to involve all water users and build a shared vision for the river. It is quite promising since cooperation at the Basin level is lengthy and can be difficult. It is usually related to the history of the region.

In Africa, there is a strong need for energy and today, the potential of hydro-electricity is not fully exploited. We know that there are limits to water infrastructures on rivers. We need to build integrated dams on the natural environment and respect the water cycle. The INBO provides support to ensure the success of the process. From our experience, you need to ensure a dialogue between stakeholders, in particular, local authorities, water users and people working with the river. We don't target the public directly, but the people who have a link with the river.

Successful communication is local rooted communication

Usually, informing and educating others on the legislation is the most difficult. At the basin scale, each country has a different understanding of water rights, different customs, and different types of institutions working for water resources. They need to understand the need to concretely integrate new standards and legislation. If you want to communicate efficiently, you need to bring local application in order to give concrete and local information so people can integrate them in their practices and their work. I actually think it applies to the general public as well. People want to have information about the cost, their house, and their city. In this regard, basin authorities need to develop their own communication plans and material. They can develop adequate information suited to their territories and to use adequate channels. For instance, in Africa, radio is particularly efficient at bringing messages and making people aware of the approach their states are taking to better the management of their river.

Many countries are very successful in bringing awareness towards the public. In Brazil and Mexico, they are advanced in bringing cooperation between water stakeholders, and involving different water users. In Mexico, the Basin Authority organized 3 to 5 days of debates and workshops to help people get to know each other and work together. They are very advanced in their use of media, and social media, in particular. In France, we are still lacking such approaches.

Many ask why the world forums are unsuccessful. I think they are a crucial place to meet other water professionals. But the forums don't communicate enough with both the local utilities and stakeholders, and the public. They are not integrating the local reality, the needs from the field, the concrete practices. Sometimes, it feels like the water professionals are living in a separate world from the public sphere, etc. Water professionals communicate on their priorities, their political messages rather than real challenges. Consequently, the public lacks adequate information. They have to understand contradictory messages especially in the media.

For instance, the media put a focus on alarmist messages such as water poisoning, death, and lack of water access, droughts or flooding. They bring negative messages over the lack of water resources. They also focus on private and public water management, etc. They argue for a privatization of resources. No countries have privatized their water except the USA. Water resources belong to utilities and many solutions are available. Today, we need to focus on the right message: to finance water infrastructures to provide water to all water users. It means the water sector needs to develop studies on water economics, water governance, and to promote a successful dialogue between users. There is a real need to better understand public perceptions and put emphasis on these processes, etc. In this sense, the basin level approach has a strong experience in promoting dialogues between stakeholders.

INBO, the International Network of Basin Organisations, aims at providing information to water stakeholders. It acts as an information player. www.inbo-news.org/

Organisation:	International Network of Basin Organisations
Creation Date:	1994
Budget:	226,000 euros
Target Group:	Water organisations
Audience:	192 members/1, 3 millions visitors (website)

8.2 INTERFACING WITH THE PUBLIC, THE ROLE PLAYED BY UTILITIES

Are water utilities the future of Water Communication? Water utilities have a strong potential to bring key messages to the public. They have the possibility to bring local messages and to organize events that can generate changes. In this frame, they are important leaders for Water Communication. They usually lack of budget to conduct efficient campaigns, they also lack of trained people or adequate resources to inform, educate and disseminate messages. Guides developed by different institutions and funding agencies confirm the lack of resources to utilities. It also confirms that many utilities might not see the importance of communicating on the value of water. Brita Forssberg highlights the impacts of long lasting strategies and communication towards the public for Stockholm based on her 20 years experiences in the organization. Patricia Bakir shows the importance of the new social media for utilities as key channels to target new publics.

8.2.1 Awareness and Small Changes Make Great Differences

By Brita Forssberg
Stockholm, Sweden

A campaign is an "intensive activity during a certain time aiming at having an influence". Public campaigns are mainstream for consumer products but for better water understanding they are rare events. A famous

Japanese professor stated that people in general have understood and accepted the energy agenda. But they have neither understood nor accepted the water agenda. Water utilities need to engage with and influence the water users. But how can they make the activities pay off? Consistent campaigns are part of the answer.

Water campaigns can reduce problems quite effectively and at long-term – even certain technical problems. Individuals who make small changes in their daily life – like reducing the amount of detergent when washing – make very significant contributions to a better and safer water environment, while adding to their own understanding of water.

A colleague of mine in another municipal administration with important customer contacts used to say that she distrusted campaigns. They were only an extrovert way of spending big money. I maintain that campaigns can have strategic importance and give extremely good, long-term results of a size, which could never be achieved by other measures at an equally low price. They can support other measures that are taken in the daily management of water utility - as well as attract new employees.

Attractive campaigns with a surprising, relevant and clear message including a human touch are seen, understood and remain remembered. They make people change their attitudes, even their behaviour. People's awareness of their own role and impact on the water environment increases. Follow-ups via personal interviews show that also their confidence in the water supplier increases. Would you believe that?

First an easy target: Washing powder

In the early 1990s the (public) Stockholm Water Company, Sweden, launched its first consumer campaign after many years of engaging with industries and many other professional activities in order to limit certain pollutants by replacing them with less harmful substances or products. The specific aim was to make the wastewater sludge fit for spreading on farmland in order to safely return nutrients like phosphorus and nitrogen as well as organic material to the soil. The monitored results were improving but still the public debate was very negative among NGOs, environmentalists, scientists, farmers and water utilities. It was time to approach private persons. What a challenge!

The decision to start a campaign (probably the first water campaign ever in the country) was based on a chemical investigation of pollutants in wastewater from households in Stockholm. The result was astonishing: over 50% of the amounts of pollutants were directly attributable to household chemicals.

Cleaning agents, solvents, paints, pharmaceutical products, odour-killing substances in socks, anti-bacterial compounds in toothpaste etc. are meant by the producers to simplify everyday life. Little attention is given to their impact on the water environment. Solid waste grows to an expensive problem when flushed into the pipe network via toilets, sinks and other outlets. All has to be handled at a cost. Awareness can save money.

Change a little. Do a little. Improve a lot.

The water utility was leading the planning process. In cooperation with a carefully selected advertising agency and in close contact with several municipal authorities as well as environmental organisations the target groups and target issues were studied and decided. The theme was simple: Use less washing detergents, your washing will be clean and you save both money and the Environment.

Financing was planned and secured, of course, as this campaign would be costly, involving three hundred thousand households via direct regular mail – not only once, direct mail to housing companies and shops, placards on buses, full page advertisements in the major daily newspaper and a press conference. Internal communications activities were also included as it was absolutely necessary that the Stockholm Water personnel were well informed and prepared. There would be questions to answer once the campaign became official and started spreading its messages about how to wash with less negative environmental impact.

The campaign got a flying start thanks to the very engaged journalists who attended the press conference and wrote about the campaign, its aims and the news that a municipal company went public in such a surprising way. The campaign was meant to last for 6 months but was actively remembered by many in Stockholm for almost 4 years, was discussed in the media and contributed to the introduction of environmentally adapted detergents on the Swedish market.

Picture campaigns in the subway keep changing minds

The following campaigns were of much lesser scale but still highly observed. They have been taking place twice a year in the subway stations during periods when the huge billboards were fairly cheap to rent for a fortnight, meeting the eyes and minds of over 2 million travellers daily. Here the messages have been presented as series of 4 attractive, surprising, warm pictures telling simple stories how to protect the environment by doing just a little. A summing up text and the company logo followed the pictures. The results were evident in the monitoring of wastewater and sludge as well as in personal interviews with Stockholmers aged 17-75.

Rebellious toilets demonstrating in the Old Town stated, "We are no dustbins!" And the amount of solid waste in the sewers declined. Wall-painting young people learned that water-soluble paints are not environmentally-friendly – and the collection centres for harmful products got more visitors. Over 65% of the travellers remembered the first subway campaigns. The 2006 campaign was the most successful: 91% observed the fish in their bowls (see Figure 8.2) and their need for protection against solvents and other harmful products. The aim is now to keep that level by repeating the important messages in smart and attractive ways. Or else they will soon be forgotten and the impact of the campaigns gets lost.

Water education in schools is another form of campaigning

Water education in different shapes has been part of strategic communications programmes and campaigns for drinking water and the protection of water catchments in several European towns and regions.

A British water communicator stated that it was nearly useless to try to inform grown-ups. They already "knew it all" and did not listen. But young people, especially the very young, were curious, understood and wanted to learn. This thinking took the form of offering free water education also in Stockholm schools. The young, dedicated specially trained water ambassadors became very popular. They visited classes and groups of teachers all over the town for many years. Their teaching made such an impression that water education is now on the school curriculum.

The water sector needs more experts!

The Dutch water sector is facing thousands of vacancies that cannot be filled if the inflow of students into water related studies does not increase. In 2008 the Dutch engineering consultancy firm DHV created a campaign **"The water sector needs YOU!"** It had the personal approach that convinced high school students that the water sector was the place to be. The website www.ikonderzoekwater.nl (istudywater) makes it possible for them to get in contact with experts from hundreds of municipalities, water boards, provinces, consultants, contractors and NGO's. The young people get training and jobs – and the water sector is able to close the personnel inflow gap! Thanks to a smart, continuous campaign that makes water interesting.

Communication – visibility - understanding

Water utilities provide the most important services of all. Still many of them are reluctant to take a visible place on the public arena and in the debate but complain that no one listens to them. Perhaps they have not said the right thing in the right way to the right people?

Water utilities have technical expertise but they have much to gain strategically from employing also other professionals - communication specialists - who can help establish long-term customer and stakeholder relations. Utilities that communicate well will be better understood and respected by society at large, by politicians, urban planners and other important stakeholders. They will be heard and listened to.

Figure 8.2 Poster of campaign by Stockholm Water Company, 2006.

8.2.2 The Role of Social Media for Utilities

Interview with
Patricia Bakir
Consultant, Jordan

In my research for the IWA European utilities conference April 2012, I focused on the use of social media by utilities. I found, online, some good examples, and good papers - but not "plenty". In the past year I'd say that many more utilities are using social media, as is the water sector overall. Based on my experience, I think DC Water is one of the most prominent in using social media.

I wouldn't begin to feel qualified to say who is the most innovative. And I'd also say that it's not the "use of social media" that's innovative; it's the content of whatever the campaign is that also matters. In other words, social media is a *tool* - but a well managed campaign depends on having your original premise correct - what is your behaviour change goal? Who are your target group(s)? What is the best way to reach each of these groups (i.e., both emotionally/message-wise, and which communication methods)?

No matter where we are in the world, word of mouth and personal contact is still the best method to persuade people to support a cause, to trust their utility, etc., etc. Social media methods built on real dialogue with stakeholders, not "providing information", will be the most successful.

Some of the best examples are actually in developing countries, where other means of communication are actually harder. I read recently about some good examples in Kenya. For example, MajiVoice (www.majivoice.com) is a service initiated by the Water Services Regulatory Board (WASREB) with World Bank support, which allows two-way dialogue with water companies. In Middle East, Miyahuna, the water company supplying water for Amman, Jordan, has adopted a system of using text messages to keep all company employees informed of any major water distribution delays (Figure 8.3). The activity is supported by the US Agency for International Development. This serves a dual purpose (1) many field employees (who do not have office e-mail/computers) are in direct contact with customers in the street, so can provide "simple" answers to customers if there's a water disruption, and (2) employees are better informed about Miyahuna news, which makes them feel a stronger part of the team and more connected to their Headquarters. Miiyahuna is also using text messaging to inform customers of the amount of bills due.

Figure 8.3 Miyahuna, staff announcement of new internal text messaging system.

My point is that whether it's using social media or general communication methods, internal communication is just as important as external - and is the job of everyone in the utility. Communication is not just the role of the communications department, and good communication starts from the top - I heard an example recently of a US company where the communications manager couldn't even do an employee newsletter of one page without detailed vetting by the CEO - the interesting thing is that sometimes social media gets around this. Once a policy has been set, then the CEO loses control whether s/he likes it or not!

Social media should be part of any overall campaign. It certainly can be a lot cheaper - but only if carefully thought out and if it does not add an additional burden that the utility cannot handle. Often social media is adopted (in all sectors) as "we have to have a Facebook page, LinkedIn, etc., etc." and then organizations cannot keep it up. So, just like any communication campaign, utilities have to determine what is most efficient in their particular situation, and make sure they can manage anything they start. Nothing is worse than a Facebook page with no updates, for example. Or a Twitter account, the same!

Today, I'll make two recommendations besides to stop reinventing the wheel, etc.

1. Use existing organizations/opportunities to share information on social media use for the water sector - such as a new LinkedIn group on Social Media for Water Utilities, and the IWA Public and Customer Communication Specialist Group Waterwiki page. The Specialist Group also has established an open Linkedin group, where we are encouraging people to share experiences.
2. Social media and its uses needs to be mainstream - not just talked about by communication specialists. Every international conference for water specialists should include a topic on this - and perhaps workshops where we can all learn from each other and from the experts outside of our sector on this. Most of us are learning by doing, the best of the companies are hiring social media specialists who then learn about water. Some of us take on this new learning better than others. We need to make it something everyone understands, everyone can "handle" (in terms of sheer volume of material/information), and everyone can participate in. Some companies (in general, I assume water organisations, too) manage to identify a "normal employee" who uses social media a lot and s/he becomes one of the "social media spokespeople" since they are already great at it. Often these are more junior staff members.

One last comment would be to deal with the managers/CEOs - some embrace this, others are "social media dinosaurs", etc. I don't feel that the water sector is "behind" other sectors, the marketing and business world is always at the forefront of any of these changes as it's the nature of private sector/entrepreneurs.

This is true of all development sectors - thus the strong emergence of social marketing, which takes its principles from business marketing. What is happening in the water sector is similar - we are learning from the private sector how to market our services, how to keep our customers happy, how to use social media as a promotional tool for the behaviour changes we want people to make - whether on the local level of conserving water, or on an international level to get governments and people to place a higher priority on water and sanitation.

9 Local-Community level: Raising public awareness

"Water is life", "Water is a right", "Water is for all", etc. Slogans about water tend to target citizens. Much emphasis today is put on users, the public, or individuals in order to bring awareness about water cycle management. Looking at scientific work, the focus is brought on:

- Understanding people's behaviours towards their daily water uses and abilities to change. Research has been focusing on public perception of water quality and drinking water (Moser, 1984; Doria, 2010) and much research can be found at the local level (at the scale of the city, river's basin or delta, state, etc., see Bibliography) or comparing the consumption habits of tap versus bottled water (Bibliography).
- Evaluating the public perceptions of risks and their environment: bathing water, pollution of natural resources, states of coastal zones, climate change, etc. (see Bibliography).
- Accompanying processes for changing behaviours and perceptions in particular to promote water conservation, and the acceptance of alternative drinking water resources in particular recycled water (Bibliography). Several are developing methodologies and studies to improve public participation to water cycle management and water programs (see Bibliography).

Today, the promotion of water cycle management is of high importance. As stated, the complexity of the water cycle governance and the multiplicitiy of discourses, messages and expertise brought confusion into the public sphere and for individuals. Citizens tend to build their opinions and behaviours based on their cultural references, peers' opinions and traditional beliefs. The need to bring education and new behaviours calls for a change when shaping public campaigns on water cycle management. It means to survey the objectives, the target and their perceptions, and the adequate channels to bring messages and changes.

Addressing public opinion on water cycle management appears to be needed at the local scale with long-term vision, different types of campaigns and clear messages. In this chapter, Karen Kun explains the need to break from traditional ways of communicating about water towards more social and marketing approaches. She shows the importance of evolving the relationship with water by taking ideas and concepts from other fields. Manjunath Sadashiva, Isaac Nyengue Bahanak and Sara Dolnicar review the necessity to address the public opinion from three different perspectives: the age e.g. targeting younger generations; the community e.g. at the scale of a village; and the topic e.g. new technologies and alternatives to drinking water. For the three experts, public perceptions and public participation are at the core of building communication processes, strategies and concrete step to advocate changes or promote new messages.

9.1 EVOLVING OUR RELATIONSHIP WITH WATER

Interview
With Karen Kun
Waterlution, Canada

The purpose of Waterlution is to inspire pattern-making and pattern-breaking change towards a healthy and sustainable relationship with water. This means the Waterlution team works to create systemic and deeply rooted change within individuals, organizations and industries understandings of their behaviours and interactions with water. Our work also bridges the gap between professionals in the water space and those who are simply interested in learning more about water. We observed that those working in the water space could be siloed in their area of specialization, potentially missing out on opportunities for collaboration. We also observed that often the messaging from those in the water field tends to be highly technical and scientific. The Waterlution approach focuses on the human connection to water and a common language to communicate this connection to entice and include people from diverse backgrounds in conversations around water. This shift in discourse represents an evolution in how individuals relate and commit to water stewardship and sustainable management. That is why we named the organization "Waterlution". Many people think the name means "water solutions", but it is actually "water and evolution" and our tagline is: evolving our relationship with water.

Waterlution encourages people to look at the significance of water in their life from a: practical use perspective, spiritual connection, and social reference to realize ways to respect this pervasive connection. Waterlution programs encourage participants to envisage water as a continuum of a personal relation with each other and our environment, creating a community and camaraderie for water's well-being. Our approach ensures that people continue to realize the inspiring power of water and give them opportunities to deepen their connection with it. It was challenging to find the right name to convey the idea of connections, relations, and deeper meanings attached to the word "water". In Canada, when you create a non-profit organization, you are obliged to choose a traditional name. You have to signify in your title what you do. In this context, if you want to work with water, you have to choose a name with...water! It limits how you can create a brand, a concept, and an approach. I'm from a marketing background and I found that these conventions stifle one's creativity. Despite that, we managed to agree that "Waterlution- A water learning experience" as a good fit to the work we were about to embark on.

A key area of focus in the Waterlution approach is making water appealing to people who initially think they are not "water people" or are interested in water. It is exhilarating to create and market programs that resonate with: individuals who are beginning to explore their interest in water and those who are more established in the water sector. Professionals in the Canadian water sector tend to focus on the technical aspects of water. So, inviting them out of their silos, exposing them to new ideas, trends and research from other sectors and offering them the space to engage in dialogue with other water professionals and non- professionals was not easy at first, until they tried it and realized the value in it. Over the last eleven years, Waterlution has gained respect for its boundary-breaking approach. To continually facilitate this cross-sector approach, I am constantly developing diverse relationships with those in and "outside" the water world. I work to gain trust with stakeholders, to involve them in our programs, to be part of a larger conversation about water, to inspire them to change their approaches and to assist people in realizing that we are ALL stakeholders in water.

Anyone can be a participant in our programs, however our main target audience is young leaders, aged 18-35 years of age. We believe this is an important demographic as their influence for positive change is significant and will grow in the future. The demographic of 18-35 year olds is characterized by adaptability, familiarity with technology and an ability to quickly learn new technologies – this is extremely powerful. If you want to capture this demographics interest and inspire them to find ways to act and bring about change, you need some magic and community. You have to bring the vision and build the right relationships for people to buy-in. This is where I think Waterlution is unique. We envisage relationships to water through long-term processes, with new ways of framing and talking about water challenges and opportunities.

My recommendation for people who want to create new campaigns around water: know about water, the discourse, and the mainstream interests, etc. And then, spend as much time as you can with people who are not in the water sector to open up your spectrum of thinking. In this process, you'll visualize your water focus from new perspectives that will help you to evolve from your initial vision, and allow you to reach your audience in new creative ways.

To find out more about Waterlution visit: www.waterlution.org

9.2 CHILDREN, AGENTS OF CHANGE, CHILDREN'S MOVEMENT FOR CIVIC AWARENESS

By Manjunath Sadashiva
CMCA, India

The Children's Movement for Civic Awareness (CMCA) is 14-year-old volunteer-driven, civil society organization that conducts a national 'civic club' program in schools across India. We define active citizenship as a process of "feeling the need for change, believing that change is possible, acting to bring about the change and influencing others to change". CMCA curriculum and methodology is designed to enable young people acquire relevant knowledge, skills and dispositions to grow up as active citizens and "Agents of change". CMCA conceptualizes active citizenship as an embodiment of four interrelated behavioural attributes of Concern–to be mindful of inconveniences and harm one's actions may cause others, to be concerned for others, environment, less privileged etc., Cooperation – civic duties such as voting, participating in community welfare activities, partnering the government, respect for rule of law and so on, Coexistence – practicing equality and celebrating diversity, and Confrontation – to complain, litigate, oppose, protest etc. to secure one's rights and of others using nonviolence methods. CMCA's unequivocal focus on nurturing active citizenship for change or societal transformation assumes critical importance in today's India that is beset with deeply entrenched patriarchy and inequality, rising intolerance and violence of all hues and colours, extreme poverty, weak public accountability, rampant corruption and human rights violations accompanied by large scale irreversible ecological damage.

Water and sanitation are two of the recurrent themes for practice related expected outcomes in CMCA's activities in addition to themes such as traffic and road safety, energy, solid waste management and greening. Water literacy, inculcating the habit of responsible use of water and rainwater harvesting are frequent foci of our school level campaigns. Sewage treatment plants and rainwater theme parks figure prominently on the field visit itinerary of our member schools. Water serves as a powerful topic for civic awareness as it best illustrates the need to strike a fine balance between claiming rights and fulfilling civic responsibilities that constitute the two sides of "active citizenship". It is easier to demonstrate the harmful impacts of irresponsible civic behaviour in the case of water in terms of depleting ground water resources, water pollution, waterborne diseases etc. and so on. At the same time, water being the life sustainer and a public good, also serves as a rallying point for rights based advocacy amongst children from less privileged communities for improved access to potable drinking water. Water also serves as a useful topic to combine popular science education and civic awareness objectives.

We organized four types of activities listed below have proven to be very effective in enhancing water awareness amongst children as compared to only classroom sessions as they promote learning by doing using a hands-on approach:

1. **Field visits** to sewage treatment plants, rain water theme parks, drinking water processing and supply plants, lakes, and water deprived communities have proven to be very effective since they entail a visual effect and offer practical exposure while also offering opportunities for children to directly

interact with field level experts and residents alike. Briefing and debriefing the children before and after the visits respectively, are necessary for enhancing effectiveness of learning.

2 **Water Clicks - Photo contests:** Children are given photo cameras and asked to click anything related to water that interests them in their everyday life. Children then select a few photos for presentation to a jury (including children and experts as members) explaining the reasons for selecting a particular situation for photo capturing and their understanding of the situation. An orientation before the photo contest and a discussion session followed by children's presentation is critical to foster.

3 **Neighbourhood /City Water Literacy Survey or Water Report Card:** children are trained to conduct a household survey of water awareness and practices in the school's neighbourhood using a simple questionnaire. Each group of 3 to 4 children is assigned 10 households/shops and each group interviews people on water literacy and household practices and water problems being faced by the households and handover a water literacy leaflet post interview to create water awareness. The data entry and analysis are centrally undertaken by the organization and the findings are publicly disseminated in the presence of children, media representatives and a Panel comprising of water supply and sewerage agency representatives, environmental activists and so on.

4 **Campus Water and Sanitation Audits**: Drinking water and sanitation is one of the key amenities covered through Participatory School Profiling (audit) exercise and has proven to be impactful in enabling improved students' access to drinking water and sanitation facilities in public schools. Students are trained to conduct the audit and share the findings with school authorities in a bid to advocate not only problem resolution but also water conservation measures such as plugging water leakage points, responsible use of water, rain water harvesting systems and so on. Campus audit not only strengthens children's critical thinking and analytical abilities but also builds their self-confidence in interacting with authority figures thereby also promoting campus democracy.

Policy domain is a complex arena where multiple interest groups compete and contest to have their interests addressed. If the policy goal were to motivate citizens for water conservation efforts or to redress water woes faced by a given community or to protect particular water bodies, then, "Wat're We Doing" video project presented as part of the 2nd French-American Science Festival in Chicago in October 2012 can indeed serve as an effective tool for policy advocacy. As such, any policy advocacy effort using videos as a tool is an effective way to attract policy makers' attention, but those featuring initiatives such as campus audit or neighbourhood/city survey of water literacy and household water usage practices [described in the preceding section] would be more effective as they tend to generate a broader view and cross sectional data that would be of more interest to policy makers than data related to a very specific situation.

If I had to provide recommendations from our leaning lessons based on CMCA's experience of working with children, communication campaigns on water addressed to children would be effective when they:

1 Offer hands-on experience based on the principle of "learning by doing"
2 Facilitate critical thinking, analysis and self reflection
3 Relate profoundly to local contexts and their everyday life situations and then zooming out to national or global contexts
4 Provide practical tips and realistic solutions adaptable to their daily routine
5 Simplify "science and technology" related to water
6 Compare examples and case studies
7 Emphasize active citizenship

CMCA's systematic and structured intervention is dedicated to moulding and nurturing active citizenship in children and youth across the country while inculcating democratic values in them. During the year

2013-14, CMCA is reaching out to 15,319 children in 7 cities and 17 villages through 295 schools, 254 volunteers and 370 civic clubs. The organization is headquartered in Bangalore and is present during the current year in 7 cities: Bangalore, Mumbai, Mysore, Hosur, Hubli-Dharwad, Nagpur, Trivandrum, and 17 villages in the Southern state of Karnataka. The member children of CMCA are in the age group of 13-14 years and represent India's diversity in terms of culture, religion, and socio-economic backgrounds. www.cmcaindia.org

9.3 ACHIEVING MDGS AT A LOCAL SCALE MEANS. PUBLIC PARTICIPATION

By Isaac Nyengue Bahanak
Junior Expert at GIZ, Cameroun

In 2007-2010, I worked for the "All for the Adoption of Environmental Actions" campaign in the Bini village in Cameroun. Bini is located 15 km from Ngaoundere (capital of the Adamawa Region) in the northern part of Cameroon. Bini Village, in the district Ngaoundere III, has been the residential centre of the University of Ngaoundere since 1982. This village, which extends over 40 square kilometres, has a an estimated population of 21,000 inhabitants. The main source of income in the village is agriculture. People grow fruit and vegetables, which they sell to students in order to send their children to school.

The aim of our campaign was to educate people about their waste management, and the management of natural resources (water and electricity). The city wanted to reduce the costs associated with the consumption of water and light in order to better the use of scarce resources. The campaign was organized by SYJEDE, a Development Support Organization created in 2004. The main objective of SYJEDE is to educate the youth and people in general on the issues of Environment and Sustainable Development.

For this campaign, we organized several actions such as a "clean-up day," which permitted the collection of nearly 150 kilograms of plastic waste in the cities of Bini, Dang and Manwoui. SYJEDE also organized an outreach visit to some industrial units in the area. Our objective during this tour in Cameroonian Society of Petroleum Deposits Ngaoundere and MAISCAM (maize processing company) was to acquire the various policies implemented by these companies in environmental management and provide some advice to various industries we encountered.

After this educational campaign, we realized that the main problem faced by the BINI population was access to drinking water. Of 100 respondents, only 20 people have access to piped water, the other 80% depend on either a well or a borehole. The indiscriminate use of these water points generates a maintenance problem. The water from these boreholes and wells, generally poorly or undeveloped, was contaminated by runoff from waste produced by the activities (laundry, dishes, cleaning, toilets, excrement). These waters were so contaminated that they caused health problems for users and a problem of pollution to groundwater. We decided to address this issue with the objective to put an end to this situation. We wanted users to take over the management and protection of these water points, so that those wells that have survived mismanagement and lack of maintenance could be saved and maintained active to accommodate the needs of the people.

We conducted a campaign of five stages:

- The first step was a detailed analysis of water samples taken from the wells and/or drilling. This step was supported by a survey of users to determine why they do not invest in the management of their water points. This step allowed us to know that users do not invest in the management of their water points because they are not aware of its usefulness. In addition, laboratory tests conducted on samples of such waters allowed us to reveal the presence of bacteria responsible for typhoid fever and a variety of strains of pathogenic microorganisms.

- The second step allowed us to select fifty wells, after a call for applications (of 81 applications received) who benefited from this project. This selection was made based on the needs expressed by the beneficiaries and depending on the budget that we had.
- The third step was the creation of management committees around these water points. These management committees are made up of users and people whose homes are located near water points. Officials were democratically appointed for each water point. Days for human investments were planned for each water point under the supervision of responsible management committees and with the technical support of the project consultants.
- The fourth step was the renovation of the proper work. The work ranged from one water point to another depending on the initial state. Thus 33 wells had their edges paved on a perimeter of a meter, 19 saw their replacement while 19 wells had high blocks to a height of 1 meter around their strings. We put an end to the use of one well because of its proximity to the sceptic tanks and the content of its water in microorganisms. As for drilling, we organized cleaning sessions around 12 holes. These sessions consisted of cleaning, weeding, painting the edges of these holes and putting plates conveying messages to draw users' awareness. However each of the water points was chemical treated using chlorine.
- The fifth and final step was a major phase of user awareness on sustainable management of this wealth.

The main result obtained after the project was the rehabilitation of 50 water points in Bini. We succeeded in gaining the public involvement for the development process of their village; the access to improved water for about 5,000 people, to streamline the management of water resources, and to decrease the number cases of diseases (related to the consumption of contaminated water) such as typhoid fever, malaria...

Access to good quality water for all is one of the eight Millennium Development Goals (MDGs) set by the UN. Achieving this goal requires the integrated management of water resources. In this context, we believed as part of the SYJEDE youth, we brought our contributions to achieving the MDGs. We leaned that with little means we can achieve many things. Through this project we received 385,000 CFA francs (590 euros) and personal contributions of the order of 400 000 CFA francs (620 euros). We significantly improved the lives of over 5,000 people.

9.4 PUBLIC PERCEPTIONS OF RECYCLED WATER AND WHY THEY MATTER A LOT!

By Sara Dolnicar[1]
The University of Queensland, Australia

Recycled water triggers negative associations in the general population (Dolnicar & Schäfer, 2009), especially among people who have no prior experience with recycled water (Dolnicar *et al.,* 2011). People are primarily concerned about their health and the health of their loved ones and the community when thinking of recycled water (Bruvold 1988; Dishman *et al.*, 1989; Higgins *et al.*, 2002; Marks *et al.*, 2002), but also think recycled water is not clear and does not taste good as shown in Table 1 which lists the top eight negative beliefs and the top eight positive beliefs expressed by the general Australian population in a survey study conducted in 2010 by Dolnicar and Hurlimann.

As a consequence of the predominantly health-related concerns of the general population relating to recycled water, people show a low willingness to use it in situations where the water comes in contact with the body, for example when drinking recycled water, showering with it or bathing in it, or indirectly

[1] Research reported in this section was funded through Australian Research Council (ARC) Discovery Grant DP0878338.

when cooking with recycled water or watering vegetables grown for the purpose of human consumption; the acceptance for uses which are far away from the human body, such as irrigation or parks and flushing the toilet, however, has consistently been reported to be very high (Bruvold & Ward, 1970; Bruvold, 1972; McKay & Hurlimann, 2003; Hurlimann, 2006; Dolnicar & Schäfer, 2006; 2009; Marks *et al.*, 2006; Po *et al.*, 2003).

Table 9.1 Beliefs people hold about recycled water.

Negative Perceptions	Positive Perceptions
May contain purified domestic wastewater	Creates new jobs
May contain purified industrial wastewater	Using it reduces the amount of wastewater discharged to the environment
May contain pathogens, such as bacteria or viruses	Is environmentally responsible
May contain trace elements of health concern, such as boron	Can save Australia from drought
Produces greenhouse emissions	Reduces the need for water restrictions
Could be a health concern, for instance if people would drink it	Increases the availability of freshwater
May contain substances such as hormones or pharmaceutically active compounds which can affect human fertility	Reduces contamination of beaches
Does not taste good	Is the most environmentally responsible water source to use

Compared to water from other sources, including desalinated water, recycled water is perceived more negatively by the majority of the population; with the exception of people who are highly concerned about the environment (Dolnicar & Schäfer, 2006; 2009). This is due to the substantial differences in people's associations with recycled water as water coming from other people's toilets as opposed to people's associations with desalinated water as clean, crisp, unpolluted ocean water (Dolnicar & Schäfer, 2009).

Figure 9.1 shows the stated willingness to use recycled water and desalinated water for a range of different purposes among Australians. Similar conclusions emerge from other studies undertaken in other geographical areas and in different years, sometimes even different decades.

Negative perceptions of recycled water and low willingness of the population to use recycled water make it difficult to implement recycling projects. A number of projects around the world had to be abandoned because of lack of public support (Hurlimann & McKay, 2004; Hurlimann & Dolnicar, 2010). In Toowoomba (Queensland, Australia), for example, the community rallied against the instruction of recycled water and successfully voted the proposed recycling project down in a referendum (Hurlimann & Dolnicar, 2010). Subsequently premiers of all other states in Australia developed desalination plants.

These historic failures and our increased understanding of how the general public perceives recycled water and why, leads to the conclusion that it is critical to manage the general public's perception of recycled water because an informed and educated population is more likely to be open-minded to adopting recycled water and is less easily influenced by scare campaigns that provide false or insufficient information to the public.

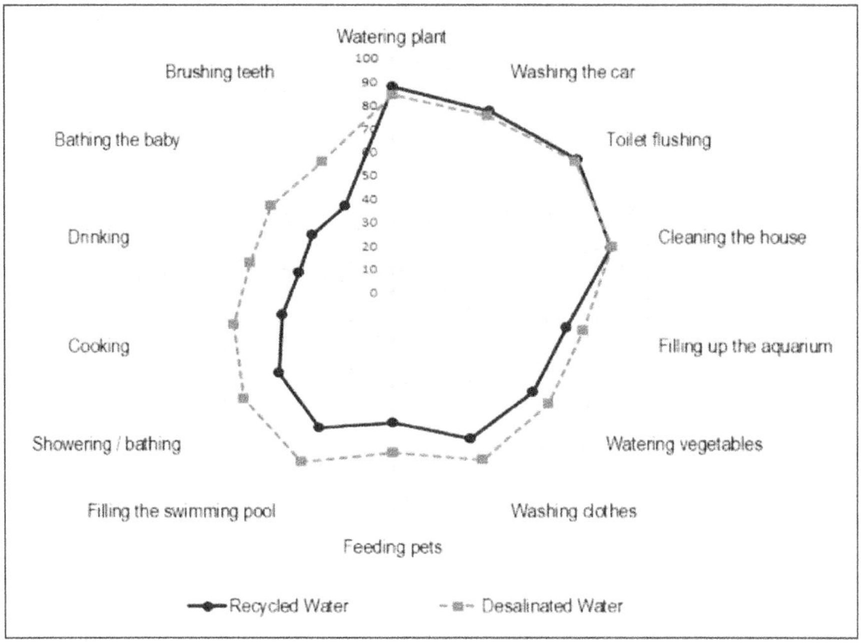

Figure 9.1 Willingness to use recycled water for different uses compared to desalinated water (2010 data collected in Australia by Dolnicar and Hurlimann).

Many avenues can be used to communicate with the public about recycled water and in so doing educate them: schools can teach students about water recycling, social marketing campaigns can be developed to explain to the general public how recycling works and what its benefits are (Dolnicar *et al.*, 2010). As shown in Table 1, not all perceptions of recycled water are negative. A number of perceptions are positive and could easily be used in communication campaigns, including the environmental benefits of recycling, the protection from drought and the reduced need for water restrictions. The media play a critical role and thus carry substantial responsibility for educating the public about recycled water (Hurlimann & Dolnicar, 2012). Other approaches of public education which have been recommended include opportunities for the general public to personally experience recycled water, such a putting "the reclaimed water in an attractive setting and invite the public to look at it, sniff it, picnic around it, fish in it, and swim in it" (Baumann & Kasperson, 1974, p. 670), the provision of simple to understand information on how recycled water is treated (Dolnicar, Hurlimann & Nghiem, 2010) and the provision of ongoing information about recycled water to the public to ensure that the population does not overreact to scare campaigns given the demonstrated recency effect of such information (Kemp, Randle, Hurlimann & Dolnicar, 2012). It is also important to note that not messages need to be conveyed by credible sources, such as people acknowledged as experts but also friends and family (Dolnicar & Hurlimann, 2009); politicians and governments – the typical conveyors of water-related policy measures - are in fact the least influential (Dolnicar & Hurlimann, 2010).

To conclude: the general public's perceptions of recycled water affect their willingness to use recycled water. It is therefore critical for all water stake holders to understand exactly how recycled water is perceived and to proactively manage these perceptions to give future recycling projects the best chance of being implemented.

10 Conclusion

So what about Water Communication?
Too much? Too few?

There is not, not enough communication about water... But, there are contradictory messages, different scales of communication and a large variety of target groups. Efficient communication on water management and resources needs to be adapted to the target public and to be implemented into sufficient materials with well established strategy. Our actual states-of-the-arts shows:

1. The most efficient and adapted communication for the water cause might be at the local scale. It supposes to involve communities and to integrate multiple criteria in particual cultural, socio-economical and political background and specificitiés. There is a strong need for utilities and water managers to further communicate to their public about water management and water resources.
2. Stereotyped messages and slogans don't serve the cause of water and our responsibility towards such an important resource is to limit such stereotypes and summarized messages. Medias play an important role and we saw how difficult for journalists the translation of water knowledge is due the rapidity of the constant evolving water terminology and translation of technical challenges in accessible terms and images. The constant evolution of wording remains related to the political agenda and international UN community which might be something to consider for future statements and events... to both avoid duplication and extensive use of new wording.
3. Water organisations should consider vocables, colors, images and their significations when setting new campaigns. We hope this book provide some useful information about setting a campaign on water, a short overview of existing messages and campaigns remain one of te first important step before entering into a communication strategy about water and by extension, sanitation.

When I had to choose the title for this book, it was very tempting to choose "Blue Washing". Catchy, new, and marketing friendly, the idea was to criticize the current mainstreaming about water management and how it brings wrong messages about water. But this book provides an overview of messages and campaigns; it wants to provide a first state-of-the-arts to show how much communication is put on water. It advocates for more studies and better integration of communication approaches and methodologies into water management.

As for Blue Washing, the idea remains. Compared to "Green Washing"[1], such "Blue Washing" is minor. Many companies are communicating about their responsibility in managing water resources and/or towards the protection of water resources. Today, there is more of blured and multiple messages about water than a strong abusive use of water as an excuse to sell or to be visible... Mainstreaming is present through some major slogans (Chapter 3). They convey contradictory messages about scarcity/flooding (climate change), insufficient water access/waste of water ressources, life/diseases and death, water wars/cooperation, peace and water shared resources... They all bring to the inner contradiction of the symbolic of water. As such, the concept of Blue Washing could be defined with the same contradictory inner characteristics:

[1]Marketing about environment engagement used to promote an organisation's products, aims and policies without actual changes in their practices and activities for the protection of te environment.

1. Marketing and campaigns used to promote an organisation's products, aims, policies and image but deceiving at the concrete level. These campaigns question legislation and general interest... How should we prevent them in the name of protecting "our water"? Should we intend to control such communication and who could?
2. All campaigns engaged to protect water according to actual needs and challenges for a batter water management and protection of water resources... Considering the difficulties of exact knowledge, Blue Washing should only serve the objective of raising visibility and responsibility towards water management and water resources.

Blue Washing implies a wider impact and engagement to coordinate and control communication about water challenges which a rather ambitious agenda. Considering how ambitious the project about setting a community and more research funds to develop Water Communication, we might want stick to developing knowledge and studies about the impacts of communication on water.

Nonetheless, we hope that the main message of this book, despite providing an overview and some guidance about communication on water, is clear: To communicate about water implies (1) analysis and well thought strategy (2) information and studies of targeted groups, materials and implementation (3) time and funds... More importantly, ethics, responsibility, and cooperation... I personally add, heart and dedication to the water cause.

References

Ader C. R. (1995). A longitudinal study of agenda setting for the issue of environmental protection. *Journalism & Mass Communication Quarterly*, **72**: 300–11.

Aim O. (2004). "Le développement durable: éléments d'analyse sur la plasticité d'un concept", Journée d'études du GRIPIC, CELSA, "Dire le développement durable", 16 juin 2004.

Ajzen I. (2002). Perceived behavioral control, self-efficacy, locus of control, and the theory of planned behaviour. *Journal of Applied Social Psychology*, **32**(4), 665–683.

Ajzen I. (1991). The theory of planned behaviour. *Organizational behavior and human decision processes*, **50**, 179–211.

Ajzen I. (1988). Attitudes, personality, and behavior. The Dorsey Press, Chicago, IL.

Agusta-Boularot S. and Paillet J. L. (1997). Le Barrage et l'Aqueduc Occidental de Glanum: le Premier Barrage-Voûte de l'Histoire des Techniques? ('The Dam and Western Aqueduct of Glanum: the Fist Arch-Dam in the History of Techniques?'). *Revue Archéologique*, **1**, pp. 27–78.

Yang H., Haynes M., Winzenread S. and Okada K. (1999). The History of Dams, UC Davis, USA.

Allély D., Drevet-Dabbous O., Etienne J., Francis J., Morel A., L'huissier A., Chappé P. and Verdelhan Cayre G. (2002). Water, gender and Sustainable Development. Lessons learnt from French co-operation in sub-Saharan Africa, Collection Etudes et Travaux, Editions du Gret, Paris, France.

Allum N. Sturgis P., Tabourazi D. and Brunton-Smith I. (2005). A meta-analysis of science knowledge and attitudes across cultures. Paper presented at the annual meetings of the American Association for the Advancement of Science.

Altheide D. L. and Snow R. P. (1991). Media Worlds in the Postjournalism Era. Aldine/de Gruyter, New York.

Anadu E. and Harding A. (2000). Risk perception and bottled water use. *Journal of the American Water Works Association*, **92**(11), 82–92.

Asch S. E. (1956). Studies of independence and conformity: I. A minority of one against a unanimous majority, in *Psychological Monographs*, **80**(9).

Atwater T. and Salwen M. B. and Anderson R. B. (1985). Media agenda setting with environmental issues. *Journalism Quarterly*, **62**(2), 393–397.

Aubry H. (2007). Imaginaires de l'eau, imaginaire du monde. La Dispute, Paris, France.

Arendt H. (1983). La condition de l'homme moderne, traduit par G. Fradier, 1983 (1961). Calmann-Lévy, Paris, France.

Bachelard G. (1942). L'eau et les rêves, essai sur l'imagination de la matière. Librairie José Corti, Paris, France.

Back W., Landa E. and Meeks L. (1995). Bottled Water, Spas, and Early Years of Water Chemistry. *Ground Water*, **33**(4).

Bakir P. (2012). Harnessing Social Media for Effective Utility Communication, Presentation, IWA World Water Congress, Busan, 2012.

Bandura A. (2001). Social Cognitive Theory of Mass Communicatio. *Media Psychology*, **3**(3), 265–99.

Baumann D. and Kasperson R. (1974). Public Acceptance of Renovated Waste Water: Myth and Reality. *Water Resources Research*, **10**(4), 667.

Berger P. L. and Luckmann T. (1967). The social construction of reality: a treatise in the sociology of knowledge, Anchor Books, Doubleday, USA.

Bielak A. T., Campbell A., Pope S., Schaefer K. and Shaxson L. (2008). From Science Communications to Knowledge Brokering: The Shift from Science Push to Policy Pull. In: Communicating Science in Social Contexts:

New models, new practices, D. Cheng, M. Claessens, T. Gascoigne, J. Metcalfe, B. Schiele and S. Shi (eds.), Springer, Dordrecht, Germany, pp. 201–226.

Bielak A. T. and Schaefer K. A. (2006). Linking Water Science to Policy: Results from a Series of National Workshops on Water. *Environmental Monitoring and Assessment*, **113**(1–3), 431–442.

Biswas A. K. (2009). Impacts of Megaconferences on Global Water Development and Management. In: Water resources Development and Management, A. K Biswas and C. Tortajada (eds.), Springer Verlag, Berlin Heidelberg Germany, pp. 3–22.

Biswas A. K. (2004). From Mar Del Plata to Kyoto: An Analysis of Global Water Policy Dialogue. *Global Environmental Change*, **14**, 81–88.

Biswas A. K. (1970). History of Hydrology. North Holland, Amsterdam and London, the Netherlands and the UK.

Bouquillon P. Matthews et J. T. (2010). Le web collaboratif - Mutations des industries de la culture et de la communication. Presses Universitaires de Grenoble, Grenoble, France.

Boure R. (2006). L'histoire des sciences de l'information et de la communication. Entre gratuité et réflexivité (1). *Questions de communication*, **10**, 277–295.

Bourg D. (2001, ed. 2003) Le Nouvel âge de l'écologie. Charles Léopold Mayer/Descartes et Cie, Paris, France.

Bourg D. (2001). Quel avenir pour le développement durable?. Le Pommier, Paris, France.

Bourg D. (1990). Les scénarios de l'écologie. Débat avec Jean-Paul Deléage. Hachette, Paris, France.

Boykoff M. T. and Timmons Roberts J. (2007). Media coverage of Climate Change: Current Trends, Strenghts, Weaknesses, Human Development Report, background paper.

Boykoff M. T. and Jules M. B. (2007). Climate Change and journalistic norms: A case-study of US mass-media coverage. *Geoforum*, **38**(6) 110–1204.

Brelet C. (2012). Anthrop'eau. L'anthropologie de l'eau racontée aux hydrologues, ingénieurs et autres professionnels de l'eau. L'Harmattan, Paris, France.

Breton P. (1996). L'argumentation dans la communication. La Découverte, Paris, France.

Brossard D. and Shanahan J. (2003). Do citizens want to have their say? Media, agricultural biotechnology, and authoritarian views of democratic processes in science. *Mass Communication*, **3**, 47–63.

Brown J. J and Reingen P. H. (1987). Social Ties and Word-of-Mouth Referral Behavior. *Journal of Consumer Research*, **14**, 250–62.

Bruvold W. H. (1991). Public Evaluation of Municipal Water Reuse Options, Technical Completi on Report, University of California, Berkeley, USA.

Bruvold W. H. and Smith B. R. (1988). Developing and assessing a model of residential water conservation". *Water Resources Bulletin*, **24**, 661–669.

Bruvold W. (1988). Public Opinion on Water Reuse Options. *Journal WPCF*, **60**(1), 45.

Bruvold W. (1972). Public Attitudes Towards Reuse of Reclaimed Water, USA, University of California.

Bruvold W. H. and Ward P. C. (1970). Public attitudes toward uses of reclaimed wastewater. *Water and Sewage Works*, **117**, 120–122.

Bryant J. and Cummins R. G. (2007). Traditions of mass media theory and research. In: Mass media effects research: Advances through meta-analysis, R. W. Preiss, B. M. Gayle, N. Burrell, M. Allen and J. Bryant (eds.), Mahwah, NJ: Lawrence Erlbaum Associates.

Burlingame G. A. Brock G. L. and Knoll L. N. (1987). Flavor Profile Analysis and Consumer Attitudes, Proceedings AWWA (American Water Works Association), Water Quality Technology Conference, USA.

Caubet C. (2002). Un tribunal de l'eau en Amérique centrale. dans *Ecologie & Politique*, **24**, 43–49.

Caulier B. (1990). L'eau et le sacré, les cultes thérapeutiques autour des fontaines en France du Moyen-âge à nos jours. Editions Beauchesne, Paris, France.

CDP (2010) CDP Water Disclosure 2010 Global Report. Carbon Disclosure Project. London. https://www.cdproject.net/CDPResults/CDP-2010-Water-Disclosure-Global-Report.pdf (Accessed 22 November 2013)

Cialdini R. B. (2001). Influence: Science and practice (4th ed.). Allyn and Bacon, New York.

Charaudeau P. (1997) Le discours d'information médiatique. La construction du miroir social. Nathan, Paris, France.

Cox J. R. (2010) Environmental Communication and the Public Sphere, (2nd ed.), Sage Publications, Thousand Oaks, CA, USA.

Coleman J. S. (1990) Foundations of social theory. Cambridge, MA: Harvard University Press, USA.

Coleman J. S. (1987) Microfoundations and macrosocial behaviour. In: The micro-macro link (pp. 153–173). J. C. Alexander, B. Giesen, R. Munch and N. J. Smelser (eds.), Berkeley: University of California Press, USA.

Coleman J. S. (1986). Social theory, social research, and a theory of action. *American Journal of Sociology*, **91**, 1309–1335.

Conseil d'État (2010). L'eau et son droit, rapport public de juin 2010, La documentation française: Paris, France.

Corbett J. B. and Durfee J. L. (2004). Testing Public (Un) Certainty of Science: Media Representations of Global Warming. *Science Communication*, **26**.

Crutchfield R. S. (1955). Conformity and character. *American Psychologist*, **10**(5), 191–198.

D'Almeida F. et C. Delporte. (2003). Histoire des médias en France, de la Grande Guerre à nos jours. Collection Champs Université, Flammarion, Paris, France.

D'Almeida F. (Dir.), (1997). La Question Médiatique: Les Enjeux Historiques et Sociaux de la Critique des Médias. Séli Arslan, Paris, France.

D'Almeida N. (2001, 2007, 2 éd.). Les promesses de la communication. Presses Universitaires de France. Paris, France.

D'Almeida N. (2007). La société du jugement. Armand Colin, Paris, France.

Dalrymple K. E., Shaw B. R., and Brossard D. (Forthcoming) Following the Leader: Using Opinion Leaders. *Environmental Strategic Communication. Journal of Society and Natural Resources.*

Deanne, J. (2005). Media, democracy and the public sphere. In Hemer, O., Tufte, T. (Ed.). Media & Glocal Change: Rethinking Communication for Development. Buenos Aires: CLASCO; Suecia: Nordicom

De Certeau M. (1988, ed. 1990) L'invention du quotidien, t. 1 Arts de faire. Gallimard, Folios Essais, Paris, France.

DEFRA (2008). Consumer attitudes to water efficiency of bathroom fittings: quantitative research: Kathryin Rathouse Social Research and Consultancy, St Albans, UK.

Delli Priscoli J., Dooge J. and Llamas R. (2004). Water and Ethics, Overview, UNESCO, Paris, France.

De Vanssay B. (2003). Les citadins et l'eau dans le monde. Représentations et comportements à Jakarta, Ouagadougou, Brasilia, Madrid, Munich et Osaka. *VertigO: La revue en sciences de l'environnement*, **4**(3), 1–6.

De Vanssay B. (2003). Les représentations de l'eau. *VertigO: La revue* électronique *en sciences de l'environnement*, Hors-série 1, décembre 2003.

Dishman C. M. Sherrard J. H. and Rebhun M. (1989). Gaining public support for direct potable water reuse. *Journal of Professional Issues in Engineering*, **115**(2), 154–161.

Dishman C. Sherrard J. and Rebhun M. (1989). Gaining Public Support for Direct Potable Water Reuse. Journal of Professional Issues. *Engineering*, **115**(2), 154.

Dolnicar S. and Hurlimann A. (2010). Australians' water conservation behaviours and attitudes. Australian. *Journal of Water Resources*, **14**(1), 43–53.

Dolnicar S. and Hurlimann A. (2010). Water Alternatives – Who and What Influences Public Acceptance?. *Journal of Public Affairs*, **11**(1), 49–59.

Dolnicar S., Hurlimann A. and Nghiem L. (2010). The Effect of Information on Public Acceptance - The Case of Water from Alternative Sources. *Journal of Environmental Management*, **91**, 1288–1293.

Dolnicar S. and Hurlimann A. (2009). Drinking Water from Alternative Water Sources: Differences in Beliefs, Social Norms and Factors of Perceived Behavioural Control Across Eight Australian locations. *Water Science & Technology*, **60**(6), 1433–1444.

Dolnicar S. Hurlimann A. and Grün B. (2011). What Effects Public Acceptance of Recycled and Desalinated Water?. *Water Research*, **45**, 933–943.

Dolnicar S. and Schäfer A. I. (2006). Public Perception of Desalinated Versus Recycled Water in Australia, CD Proceedings of the AWWA Desalination Symposium.

Dolnicar S. and Schäfer A. I. (2009). Desalinated Versus Recycled Water — Public Perceptions and Profiles of the Accepters. *Journal of Environmental Management*, **90**, pp. 888–900.

Doria M. F. (2010). Factors influencing public perception of drinking water quality. *Water Policy*, **12**(1), 1–19.

Doria M. F., Pidgeon N. and Hunter P. R. (2009). Perceptions of drinking water quality and risk and its effect on behavior: A cross-national study. *Science of the Total Environment*, **407**(21), 5455–5464.

Doria M. F. Pidgeon N. and Haynes K. (2006). Communication of tap-water risks: challenges and opportunities. In: Water Contamination Emergencies: Enhancing our Response, K. C. Thompson and J. Gray, (eds.), Royal Society of Chemistry, London, UK, pp. 165–171.

Doria M. F. (2006). Bottled water versus tap water: understanding consumers' preferences. *Journal of Water Health*, **4**(2), 271–6.

Doria M. F., Pidgeon N. and Hunter P. (2005). Perception of tap water risks and quality: A structural equation model approach. *Water Science and Technology*, **52**(8), 143–149.

Driedger S. M. (2007). Risk and the media: a comparison of print and televised news stories of a Canadian drinking water risk event. *Risk Analysis*, **27**(3): 775–86.

Embid A. (2003). The transfer from the Ebro basin to the Mediterranean basins as a decision of the 2001 National Hydrological Plan: the main problems posed. *International Journal of Water Resources Development*, **19**, 399–411.

Falconer I., Bartram J., Chorus I., Kuiper Goodman T., Utkilen H., Burch M. and Codd G. A. (1999). Safe Levels and Safe Practices. In: Chorus I., Bartram J., editors. Toxic Cyanobacteria in Water. A Guide to their Public Health Consequences, Monitoring and Management. E & FN Spon on behalf of WHO; London http://www.who.int/water_sanitation_health/resourcesquality/toxcyanbegin.pdf (accessed 20th November 2013)

FAO (2011). FAO expert consultation. Communication for Development: Metting today's agriculture and rural development challenges, Office of Knowledge exchange, Research and Extenstion – Research and Extension Branch, Rome, Italy.

Feitelson E. (2002). Implications of shifts in the Israeli water discourse for Israeli-Palestinian water negotiations. *Political geography*, **21**(3), 293–318.

Ferrier C. (2002). Bottled Water: A Social Phenomenon. *A Journal of the Human Environment*, **30**(2): 118–119.

Flichy P. (1991). Une Histoire de la Communication Moderne: Espace Public et Vie Privée. La Découverte, Paris, France.

Fishbein M. von Haeften I. and Appleyard J. (2001). The role of theory in developing effective interventions: Implications from Project SAFER. *Psychology, Health & Medicine*, **6**, 223–238.

Fishbein M. and Ajzen I. (1975). Belief, attitude, intention, and behavior: An introduction to theory and research, Reading, MA: Addison-Wesley.

Foucault M. (1977). Discipline and Punish: The Birth of the Prison. Pantheon Books, New York, USA.

Freemand R. and Phillips R. (2002). Stakeholder Theory: A Libertarian Defense. *Business Ethics Quaterly*, **12**(3) (July 2002), 331–49.

Friedman S. (1990). Two decades of the environmental beat. E. E. Dennis and C. L. LaMay (Eds.) Covering the Environment published as Gannett Center journal, **4**(3), 13–23.

Georgakakis D. et J.-M. Utard (2001). Science des médias: jalons pour une histoire politique. L'Harmattan, Paris, France.

Glasser R., Manty D. and Nehman G. (1975). Public participation in water resource planning. International Water Resources Association and UNESCO Seminar, Paris and Strasbourg.

Gleick P. (2011). Bottled and Sold: The Story Behind Our Obsession with Bottled Water. Island Press, Washington D. C., USA.

Global Water Partnership (2011). Social Equity and Integrated Water Management, Humbert Pena Stockholm, Sweden.

Gobin C. and Deroubaix et J.-C. (2010). L'analyse du discours des organisations internationales. Un vaste champ encore peu exploré. *Mots*. Les langages du politique, **94**.

Goffman E. (1959). The Presentation of Self in Everyday Life, University of Edinburgh Social Sciences Research Centre, Anchor Books, USA.

Goldman M. (2005). Imperial Nature. Yale University Press, New Haven.

Goubert J.-P. (2002). L'eau, puissance civilisatrice. CIEAU, Paris, France.

Goubert J.-P. (1986). La conquête de l'eau. L'avènement de la santé à l'âge industriel. Robert Laffont, Paris, France.

Granovetter M. S. (1973). The Strength of Weak Ties. *The American Journal of Sociology*, **78**(6), 1360–1380.

Greimas J. (1966). Sémantique structurale: recherche et méthode. Larousse, Paris, France.

Gritti J. (2001). L'eau, mythes et symboliques. CIEAU, Paris, France.

Gudykunst W. B. (1995). Building bridges: Interpersonal skills for a changing world. Houghton Mifflin, Boston, USA.

Habermas J. (1962, trans. 1988) L'espace public. Archéologie de la publicité comme dimension constitutive de la société bourgeoise. Payot, Paris, France.

Herman E. S., and Chomsky N. (1988). Manufacturing Consent: The Political Economy of the Mass Media. Pantheon Books, Toronto, Canada.

Hervé-Bazin C. (2012). Eau, Femme et Développement Durable. Sarrebruck, Allemagne: Editions Universitaires Européennes.

Higgins J., Warnken J., Sherman P. and Teasdale P. (2002). Surveys of Users and Providers of Recycled Water: Quality Concerns and Directions for Applied Research. *Water research*, **36**, 5045.

Holbert L. R. and Kwak N. and Shah D. V. (2003). Environmental concern, patterns of television viewing and pro-environmental behaviors: Integrating models of media consumption and effects. *Journal of Broadcasting and Electronic Media*, **47**(2), 177–196.

Howarth D. and Butler S. (2004). Communicating water conservation: How can the public be engaged?. *Water Supply*, **4**(3), 33–44.

Hurlimann A. and Dolnicar S. (2012). Newspaper coverage of water issues in Australia. *Water Research*, **46**, 6497–6507.

Hurlimann A. and Dolnicar S. (2011). Voluntary relocation - an exploration of Australian attitudes in the context of drought, recycled and desalinated water. *Global Environmental Change*, **21**(3), 1084–1094.

Hurlimann A. and Dolnicar S. (2010). Acceptance of water alternatives in Australia. *Water Science and Technology*, **61**(8), 2137–2142.

Hurlimann A. and Dolnicar S. (2010). When Public Opposition Defeats Alternative Water Projects - the Case of Toowoomba Australia. *Water Research*, **44**, 287–297.

Hurlimann A., Dolnicar S. and Meyer P. (2009). Understanding behaviour to inform water supply management in developed nations - A review of literature, conceptual model and research agenda. *Journal of Environmental Management*, **91**(1), 47–56.

Hurlimann A. (2008). Community Attitudes to Recycled Water Use: an Urban Australian Case Study – Part 2, The University of Melbourne Research Report No 56 CRC for Water Quality and Treatment.

Hurlimann A. (2006). Melbourne Office Worker Attitudes to Recycled Water Use. *Water, Journal of the Australian Water Association*, **33**(7), 58–65.

Hurlimann A. (2006). Water, Water, Everywhere - Which Drop Should be Drunk?. *Urban Policy and Research*, **24**(3), 303–305.

Hurlimann A. and McKay J. (2004). Attitudes to Reclaimed Water for Domestic Use: Part 2. Trust. *Water, Journal of the Australian Water Association*, **31**(5), 40–45.

International Union for Conservation of Nature (2010). Water for Business: Initiatives guiding sustainable water management in the private sector, http://www.wbcsd.org/web/water4business.pdf (accessed 15-08-2013)

IFOP (2011). Les entreprises et l'eau: perceptions, comportements et gestion, IFOP pour Hydrelis, Paris, France.

Jairath J. (2003). Drought of Sanity and Flood of the Absurd: Politics of Water Discourse. *Economic and political weekly*, **38**, 665–668.

Jeanneney J.-N. (1996). Une histoire des médias, des origines à nos jours. Seuil, Paris, France.

Jollivet M. (2001). Le développement durable, de l'utopie au concept: de nouveaux chantiers pour la recherché. Elsevier, Paris, France.

Jones A. Q., Deweyb C. E., Doréb C. K., Shannon E., Majowiczb C., McEwenb M. A., Waltner-Toewsb D., Hensondand S. J. and Mathewse E. (2007). A qualitative exploration of the public perception of municipal drinking water. *Water Policy*, **9**(4), 425–438.

Karembu M. and Kiai W. (1999). Understanding Environment Communication, Oakland Media Services Limited, Nairobi.

Katz E. and Lazarsfeld P. F. (1955). Personal Influence. Free Press: Glencoe, Illinois, USA.

Keller E. B. and Berry J. L. (2003). The influentials: One American in ten tells the other nine how to vote, where to eat, and what to buy. Simon & Schuster: New York, USA.

Kemp B. J. (1989). Ancient Egypt: anatomy of a civilization. Routledge, London, UK.

Kemp B., Randle M., Hurlimann A. and Dolnicar S. (2012). Community Acceptance of Recycled Water – Can We Inoculate the Public Against Scare Campaigns? *Journal of Public Affairs*, **12**(4), 337–346.

Kirilenko A., et al. (2012). Computer-assisted analysis of public discourse: a case study of the precautionary principle in the US and UK press. *Quality and Quantity*, **46**(2), 501–522.

Knapp M. L. and Miller G. R. (1994). Handbook of Interpersonal Communication. Sage Publications, Thousand Oaks, USA.

Klofstad C. A. (2007) Talk leads to recruitment: How discussions about politics and current events increase civic participation. *Political Research Quarterly,* **60**, 180–191.

Lakin J. L and Chartrand T. L. (2003). Using nonconscious behavioral mimicry to create affiliation and rapport. *Psychological Science*, **14**(4), 334–339.

Lasswell H. D. (1927). Propaganda Technique in the World War. Peter Smith, New York, USA.

Lasswell H. D. (1948). Power and Personality. W. W. Norton & Company, Inc, New York, USA.

Lazarsfeld P. (1944). The People's Choice: How the voter make his mind in a presidential campaign. Columbia University Press, USA.

Levi E. (1995). The science of water: the foundation of modern hydraulics, translated by D. Medina, American Society of Civil Engineering, USA.

Libaert T. (1992). La communication verte. Editions Liaisons, Paris, France.

Libaert T. (2010). Communication et environnement, le pacte impossible. PUF, Paris, France.

Mackie D. M. (1987). Systematic and nonsystematic processing of majority and minority persuasive communications. *Journal of Personality and Social Psychology*, **53**(1), 41–52.

MacPherson L. (2011). Terminology, Images and Context: Key to a Sustainable Future, CH2M Hill, presentation, 24 September 2011.

MacPherson L. and Slovic P. (2010). Talking about Water. Vocabulary and Images that Support Informed Decisions about Water Recycling and Desalination, WateReuse Research Foundation Alexandria, USA.

Maganda C. (2010), Water Management Practices on Trial: The Tribunal Latinoamericano del Agua and the Creation of the Public Space for Social Participation in Water Politics. In: Social Participation in Water Governance and Management. K. A. Berry and E. Mollard (eds.), Critical and Global Participatives, Earthscan, London, UK.

Maibach E. W., Roser-Renouf C. and Taylor M. (2008). What are Americans thinking and doing about global warming: Results of a national household survey. Fairfax, VA: George Mason University, USA.

Malissar A. (2002). Les Romains et l'eau. Les Belles Lettres, Paris, France.

Mandressi R. (2009). Figures de la preuve. *Communications*, **84**, Mai 2009.

Marks J., Martin B. and Zadoroznyj M. (2006). Acceptance of Water Recycling In Australia: National Baseline Data. *Water Journal of the Australian Water Association*, **33**(2), 151.

Marks J., Cromar N., Howard F., Oemcke D. and Zadoroznyj M. (2002). Community Experience and Perceptions of Water Reuse, IWA World Water Congress, Melbourne, Australia.

Mattelard A. (1994). L'invention de la communication. Découverte, Paris, France.

Mattelard A. and Matterlard M. (1995). Histoire des théories de la communication, La Découverte, Paris, France.

McKay J. and Hurlimann A. (2003). Attitudes to Reclaimed Water for Domestic Use: Part 1. Age. *Water, Journal of the Australian Water Association*, **30**(5), 45–49.

McQuail D. (2005). McQuail's Mass Communication Theory. Sage Publications, Thousand Oaks, CA, USA.

Miller J. D. and Panach M. A. (2009). Sources and Framing in Print News coverage of a Water Quality Dispute in Oklahoma and Arkansas. *Research Paper*, USA.

Molinaro H. (2003). Plastic bottles go to waste, says Calif. Conservation agency. *Society of Plastics Engineers,* Inc., n 59 - 7: 64(1).

Morelle J.-P. (2007). Aquarelle, L'eau créatrice, Éditions Fleurus, Paris, France

Moser G. (1984). Water quality perception: A dynamic evaluation. *Journal of Environmental Psychology*, **4**(3), 201–210.

Moser G. Raitu E., and De Vanssay B. (2005). Thinking about water. Social representations, ideologies and conducts: A model of relationships with water in different social contexts). *Trayectorias: Revista de Ciencias Sociales de la Universidad de Nuevo León*, Año VII, N 18, Mayo-agosto, 79–91.

Mount D. C. and Bielak A. (2011). Deep Words, Shallow Words: An Initial Analysis of Water Discourse in Four Decades of UN Declarations, UNU-INWEH, Hamilton, Ontario, Canada.

National Film Board of Canada and Primiative Entertainment inc. (2010).
Nisbet M. C. and Kotcher J. E. (2009). A two-step flow of influence?: Opinion-leader campaigns on climate change. *Science Communication*, **30**, 328–354.
Nisbet M. C. and Scheufele D. A. (2007). The future of public engagement. *The Scientist*, **28**(10), 38–44.
Nisbet M. C. and Myers T. (2007). Twenty years of public opinion about global warming. *Public Opinion Quarterly*, **71**(3), 444–470.
Nisbet M. C. and Goidel R. K. (2007). Understanding Citizen Perceptions of Science Controversy: Bridging the ethnographic-survey divide, in *Public Understanding of Science*, **16**, 421–440.
Olson E. D. (1999). Bottled Water: Pure Drink or Pure Hype? Natural Resource Defense Council.
Onufrak S. J., Park S., Sharkey J. R. and Sherry B. (2012). The relationship of perceptions of tap water safety with intake of sugar-sweetened beverages and plain water among US adults. *Public Health Nutrition*, October, **26**, 1–7.
Opinion Leader (2006). Using Water Wisely, Consumer Council for water, London, UK.
Ostman R. E. and Parker J. L (1987). Impact of education, age, newspapers, and television on environmental knowledge, concerns, and behaviors. *Journal of Environmental Education*, **19**(1), 3–9.
Pastoureau M. (1992). Dictionnaire des couleurs de notre temps. Edition Bonneton, Paris, France.
Pearse P. H., Bertrand F. and MacLaren J. W. (1985). Currents of Change. Final Report, Inquiry on Federal Water Policy, Environment Canada, Canada.
Pendleton L., Martin N. and Webster D. G. (2001). Public Perceptions of Environmental Quality: A Survey Study of Beach Use and Perceptions in Los Angeles County. *Marine Pollution Bulletin*, **42**, 1155–1160.
Pottinger L. (2013). The Tale of Two Dams: Comparing Ethiopia's Grand Renaissance to Hoover, www.internationalriver.org
Priest S. H. (2001). A Grain of Truth: The Media, The Public, and Biotechnology. Rowman & Littlefield Publishers, Inc.: Lanham, MD, USA.
Putnam R. D. (2000). Bowling Alone: The collapse and revival of American community. Simon and Schuster, NewYork, USA.
Quoting Raucher R. S. (2005). Communicating Value: How the Public Can be Informed Effectively About the True Value of Water Service Relative to the Costs Invovled to Treat and Distribute the Water. Presented at the Thrird National Drinking Water Symposium in Denver, Colorado, October 2005.
Raucher R. S., Chapman D., Genderson J., Hagenstad M., Rice J., Goldstein J., Huber-Lee A., DeOreo W., Mayer P., Hurd B., Linsky R., Means E. and Renwick M. (2005). The Value of Water: Concepts, Estimates, and Applications for Water Managers, AwwaRF, Denver, Colo., USA.
Raynal C. (2004). La vente des eaux minerals par les pharmaciens, dans *Revue d'histoire de la pharmacie*, **92**(334), 587–606.
Rebillard F. (2009). Web 2.0 en perspective. Une analyse socio-économique de l'Internet. L'Harmattan, Paris, France. http://dwi.defra.gov.uk/research/completed-research/reports/dwi70-2-221.pdf (accessed 15-10-2013)
Robbins P. T. (2003). Transnational Corporations and the Discourse of Water Privatization. *Journal of international development*, **15**(8), 1073–1082.
Rogers E. M. (1993). Looking back, looking forward: a century of communicationresearch. In: Beyond Agendas: New directions in communication research. P. Gaunt (ed.), Greenwood, New Haven, CT.
Saad L. (2011). Water issues worry Americans most, global warming least. Gallup.
Sadler B. S. (1987). Communication strategies for heightening awareness of water in Report 2 of IHP II Project Cl (on heightening awareness of the socio-economic role of water), UNESCO, Paris, France, http://hydrologie.org/ (accessed 20-07-2013).
SAM (2007). Water, a market of the future, Sustainable Asset Management AG. Zürich, Switzerland.
Shah D. V. (1998). Civic engagement, interpersonal trust, and television use: An individual level assessment of social capital. *Political Psychology*, **19**, 469–49.
Shannon C. E. (1948). "A mathematical theory of communication." *Bell System Technical Journal*, **27**(July and October), 379–423 and 623–656.
Shannon C. E. and Weaver W. (1963). The Mathematical Theory of Communication. Univ. of Illinois Press: Chicago, USA.

Shaxson, Louise with Alex T. Bielak, et al. 2012. Expanding our understanding of K*(KT, KE, KTT, KMb, KB, KM, etc.) A concept paper emerging from the K* conference held in Hamilton, Ontario, Canada, April 2012. UNU-INWEH, Hamilton, ON. 30pp + appendices http://inweh.unu.edu/wp-content/uploads/2013/05/KStar_ConceptPaper_FINAL_Oct29_WEBsmaller.pdf Accessed 23-10-2013.

Sheikheldin G., Krantzberg G. and Schaefer K. (2010). Science-Seeking Behaviour of Conservation Authorities in Ontario. Environmental Management **45**, 912–921. (or see earlier synopsis at http://ec.gc.ca/scitech/default.asp?lang=En&n=BCABA058-1 (Accessed 15-10-2013.)

Simpson J. and Stratton H. (2011). Talking about water: words and image that enhance understanding. *Waterlines Report Series* **49**, May 2011, National Water Commission, Commonwealth of Australia, Cranberra, Australia.

Sironneau J. (2002). Le droit international de l'eau existe-t-il? Evolutions et perspectives, Direction de l'eau, Ministère de l'écologie et du développement durable, France.

Sloan W. D. (1991). Perspectives on Mass Communication History. Laurence Erlbaum Association, Mahwah, NJ, USA.

Snellen W. B. and Schrevel A. (2004). IWRM: for sustainable use of water, 50 years of international experience with the concept of integrated water management, Ministry of Agriculture, Nature and Food Quality, The Netherlands.

Smets H. (2007). La reconnaissance officielle du droit à l'eau en France et à l'international, AFD, France.

Smets H. (2010). La mise en œuvre du droit à l'eau, les solutions à Paris, Académie de l'eau, Paris, France.

Stamm K. R., Clark F. and Reynolds P. (2000). Mass communication and public understanding of environmental problems: The case of global warming. *Public Understanding of Science*, **9**, 219–237.

Stamm K. R. (1985). Newspaper Use and Community Ties: Toward a dynamic Theory. Ablex Publishing, New York, USA.

Starr P. (2004). The Creation of the media: Political origins of modern communications. Basic Books, New York, NY, USA.

Strang V. (2004). The Meaning of Water. Berg: Oxford Internation Publishers Ltd, New York, USA.

Swift M. (2008). Cartes du Monde à travers l'histoire, Geo Histoire, Les éditions Prisma/Les éditions de l'Après-Midi.

Talatala S. (2008). The Effect of Tap Water Perception on the Consumption of Bottled Water, [Online Document], May 12 2008, www.nature.berkeley.edu

Tarter J. (2005). Water Wars in Environmental Justice Discourse: Place and Identity in the Age of Globalization. *Tamkang review*, **34**(3/4), 85–112.

Terry D., Hogg M. and White K. (1999). The theory of planned behaviour: Self-identity, social identity and group norms. *British Journal of Social Psychology*, **38**, 225–44.

Tilt B. and Xiao Q. (2010). Media coverage of environmental pollution in the People's Republic of China: responsibility, cover-up and state control. *Media, Culture & Society*. SAGE Publications (Los Angeles, London, NewDelhi and Singapore), **32**(2): 225–245.

Tournier M. (1985). Texte "propagandiste" et cooccurrences. Hypothèses et méthodes pour l'étude de la sloganisation, dans *Mots*, **11**, 155–187.

Tröger K. (2010). Clés méthodologiques de la socio-anthropologie de l'environnement. Comment aborder la dimension sociale de l'eau, Agence de l'eau, France.

UNICEF (1999). A Manual on Communication for Water Supply and Environmental Sanitation Programmes, New York, USA. http://www.unicef.org/wash/files/com_e.pdf (accessed 15-10-2013)

United Nations (1974). General Assembly Resolution 3201 (S-VI): Declaration on the Establishment of a New International Economic Order

United Nations (1991). General Assembly Resolution Women, environment, population and Sustainable Development, n A/RES/46/167, 18 November 1991.

UNEP (2011). Water. Investing in natural capital, Towards a Green Economy, Chapter coordinated by Young M. D., Paris, France.

United States. Department of State (1956). The Suez Canal Problem, 26 July-22 September 1956: a documentary publication. U.S. Department of State Publication N 6392 (Washington: G.P.O., 1956), pp. 345–351.

Uslaner E. (1998). Social capital, television, and the mean world: Trust, optimism and civic participation. *Political Psychology*, **19**, 441–467.

Van Tubergen A. and van der Linden S. (2002). A brief history of spa therapy. *Annals of the Rheumatic Diseases*, **61**(3), 273–375

Van Wijk-Sijbesma C. (1979). The use of aids in environmental educational sanitation education. *Comunity Water supply and sanitation in developing countries*, January 1980, **1**(1) published with the IRC newsletter, 1980.

Velayutham C. and Arul Aram I. (2010). Tracing the Life Cycle of Water Conflict News Coverage: A Case study of Cauvery River Tribunal Final Verdict. *International Journal of Communication Studies*, **2**, June 2010.

Wagner W., Kronberger N. and Seifert F. (2002). Collective symbolic coping with new technology: Knowledge, images and public discourse, in *British Journal of Social Psychology*, **42**, 323–343.

Ward S. and Potter A. (2009). H2O: Just add Branding: Producing high-quality Children's TV Drama for multi-channel environments. *Media International Australia*, **133**, 31–42

WaterLife http://waterlife.nfb.ca/ (accessed 02-09-2013)

Watzlawick P., Beavin J. H. and Jackson D. D. (1967). Pragmatics of human communication: A study of interactional patterns, pathologies, and paradoxes. Norton, New York.

Westley B. and MacLean M. (1957). A conceptual model for mass communication research. *Journalism Quarterly*, **34**, 31–38.

Whiteford L. (2005). Casualities in the Globalization of Water: a Moral Economy of Health Perspective. Whiteford L. and S. Whiteford (2005) Globalization, Water, and Health: Resource Management in *Times of Scarcity, School of American Research Press*, Santa Fe, New Mexico, USA, 25–44.

Wilk R. (2006). Bottled Water: The pure commodity in the age of branding. *Journal of Consumer Culture*, **6**(3), 303–325.

Willner S. (2006). Strategic Communication for Sustainable Development, a conceptual overview. GTZ: Eschborn, Germany.

WMO (1992). The Dublin Statement on Water and Sustainable Development. International Conference on Water and the Environment, Dublin 26–31 January 1992, World Meteorological Organization, Geneva. http://www.wmo.int/pages/prog/hwrp/documents/english/icwedece.html (accessed 25th November 2013)

Bibliography

Adam B. and Carter C. (eds.). Environmental Risks and the Media, Routledge, London, UK, pp. 1–26.

Adams R. L. A. (1973). "Uncertainty in nature, cognitive dissonance, and the perceptual distortion of environmental information". *Economic Geography*, **49**, 287–297.

Adorno T. W. and Horkheimer M. (1944, tr. 1974). La dialectique de la raison, trad. Eliane Kaufholz. Gallimard, Paris, France.

Agence France Presse. (2003). Singapore wages high-tech campaign to promote recycled water, Retrieved June 10, 2003 from Dialog NewsRoom database.

Affset. (2005). "Perception du risque et participation du public". Santé et environnement, enjeux et clés de lecture, Paris, France.

Aitken C., McMahon T. A., Wearing A. J. and Finlayson B. L. (1994). "Residential water use: predicting and reducing consumption". *Journal of Applied Social Psychology*, **24**, 136–158.

Albro R. (2000). "The water is ours, Carajo!: Deep citizenship in Bolivia's water war". In: Social Movements: An anthropological Reader, J. Nash, (ed.)., Blackwell Publishing Ltd, Malden, MA, USA, pp. 249–271.

Alhumoud J., Behbehani H. and Abdullah T. T. (2003). "Wastewater Reuse". *Kuwait. Environmentalist*, **23**(2), 117.

Allan J. A. (2005). "Water in the Environment/Socio-Economic Development Discourse: Sustainability, Changing Management Paradigms and Policy Responses in a Global System". *Government and opposition*, **40**(2), 181–199.

Alley K. (2002). On the Banks of the Ganges: when wastewater meets a sacred river. University of Michigan Press, USA.

Altschull J. H. (1990). From Milton To Mcluhan: The Ideas Behind American Journalism. Allyn & Bacon, Boston, USA.

Anand P. B. (2004). "Water and Identity – An Analysis of the Cauvery River Water Dispute", Bradford Centre for International Development, University of Bradford, UK.

Anand P. B. (2007). "Capability, Sustainability, and Collective Action: An Examination of a River Water Dispute". *Journal of Human Development and Capabilities*, **8**(1), 109–132.

Anderson A. (1997). Media, Culture, And The Environment. *New Brunswick*, Rutgers University Press, New Jersey, USA.

Andria Q. J., Catherine E. D., Kathryn D., et al. (2006). Public perception of drinking water: a postal survey of residents with private water supplies. *BMC Public Health*, **6**, 94–105.

Arbuès F., Garcia-Valinas M.-A. and Martinez-Espineira R. (2003). Estimation of residential water demand: a state-of-the-art review. *Journal of Socio-Economy*, **32**(1), 81–102.

Aspe C. (eds.). (1991). Chercheurs d'eau en Méditerranée. Pratiques et Représentations de l'eau dans l'espace Méditerranéen, Editions du Félin, Paris, France.

Aspe C. et Point P. (1999). L'eau en Représentations. Gestion des Milieux Aquatiques et Représentations Sociales. GIP Hydrosystèmes, Cemagref, Ministère de l'Aménagement du Territoire et de l'Environnement, Paris, France.

Asthana V. (2011). "The urban water reform project: a critical discourse analysis of the water policy making process in Delhi, India". *Water policy: official journal of the World Water Council*, **13**(6), 769–781.

Aubry H. (2012). Les autres voix de l'eau: les enjeux du forum alternatif mondial de l'eau. La Dispute, Paris, France.

Auslander B. A. and Langlois, P. H. (1993). "Toronto tap water: perception of its quality and use of alternatives". *Canadian Journal of Public Health*, **84**(2), 99–102.

Baggett S., Jeffrey P. and Jefferson B. (2004). "Participatory water reuse planning: a conceptual model based on social learning and personal constructs", Presented at IWA World Water Congress, Marrakech 19–24 September, CDROM.

Baggett S., Jeffrey P. and Jefferson B. (2006). "Risk Perception in Participatory Planning for Water Reuse". *Desalination*, **187**, 149–158.

Baird J. (2013). "Public perceptions regarding water quality and attitudes towards water conservation in the lower Fraser Basin". *Water Resources Research*, **34**(5), 1299–1306.

Baker L. A., Westerhoff P. and Sommerfeld M. (2006). "An adaptive management strategy using multiple barriers to control taste and odor problems in the metro-Phoenix water supply". *Journal of the American Water Works Association*, **98**(6), 113–126.

Ballut C. and Fournier P. (2013). Au fil de l'eau. Ressources, risques et gestion du Néolithique à nos jours. Presses Universitaires Blaise Pascal, Collection Croisée des SHS, France.

Barbier F. et Bertho Lavenir C. (1996). Histoire des Médias: de Diderot à Internet. Armand Colin, Paris, France.

Barlow M. (2009). Blue Covenant: The Global Water Crisis and the Coming Battle for the Right to Water. New Press, New York, USA.

Barlow M., and Clarke T. (2005). Blue Gold. The Fight to Stop the Corporate Theft of the World's Water. New Press, New York, USA.

Baron C. (2007). De l'eau sacrée à l'eau marchandise. Représentations de l'eau en Afrique. In: Imaginaires de l'eau, Imaginaire du Monde. Dix regards sur l'eau et sa symbolique dans les sociétés humaines, H. Aubry, La Dispute, Paris, France, 109–137.

Barr S. (2003). "Strategies for sustainability: citizens and responsible environmental behaviour". Area, **35**(3), 227–240.

Barraque B. (2012). Urban water conflicts. Urban Water Series, UNESCO-IHP.

Barraqué B. (2003). Past and future sustainability of water policies in Europe. *Natural Resources Forum*, **27**, 200– 211.

Barraque B. (1992). "Water management in Europe: beyond the privatization debate". *Flux*, **7**, 7–26.

Barraqué B. et Bley D., Boëtsch G., Masali M., Rabino Massa E. (1995). L'homme et le Lac: Usages et Représentations de l'espace lacustre, VIèmes Journées de la Société d'Ecologie Humaine, 13–15 juin 1994, Cadenabbia-Griante, Italie, Supplément de l'Ecologie Humaine, 13.

Barthes R. (1970). Mythologies. Seuil, Paris, France.

BBC Research and Consulting (2007). Water Conservation Awareness, Attitudes, and Behaviors – Draft Report. Arizona Municipal Water Users Association. Phoenix, Arizona, USA.

Beaudet F. (2012). L'eau, etc. code rouge. Editions Amalthée, Nantes, France.

Beck U. (1992). Risk Society: Towards a New Modernity. Sage, London, UK.

Bell A. (1994). "Climate of Opinion: Public and Media Discourse on the Global Environment". *Discourse and Society*, **5**(1), 33–64.

Bell A. (1994). "Media (Mis) Communication on the Science of Climate Change". *Public Understanding of Science*, **3**, 259–275.

Ben Abdallah M. B. A. (1996). L'eau dans la pensée musulmane et littérature arabe. Ministère des Habous et des Affaires Islamiques, Rabat, Maroc.

Bennett W. L. (2002). News: The Politics of Illusion. Longman, New York, USA.

Berger C. et Roques J.-L. (2005). L'Eau comme fait social. Transparence et opacité dans la gestion locale de l'eau, L'Harmattan, Paris, France.

Bernadis M.-A. et Mesteroff A. (1990). Le Grand Livre de l'eau. La Manufacture, Paris, France.

Bernauer T. (2002). "Explaining success and failure in international river management". *Aquatic Science*, **64**, 1–19.

Beveridge R. (2010). "A 'politics of inevitability': The privatisation of the Berlin Water Company, the global city discourse and governance in 1990s Berlin". University of Newcastle Upon Tyne, VS Verlag für Sozialwissenschaften, Heidelberg, Germany.

Biagi M. and Ferro M. (2011). "Ecological Citizenship and Social Representation of Water Case Study in Two Argentine Cities" in SAGE Open, published online.

Biagi M. (2010). "El agua en el medio urbano de Argentina: la brecha entre pensar y hacer. II Taller Internacional sobre Sinergias Ambientales entre Aguas Continentales y las Marinas". Environmental synergies, Mnemosyne, Buenos Aires, Argentina.

Biagi M. (2006). "La representación social del agua en el medio urbano de Argentina" [Social representation of water in sociological perspective]. In: Sinergias Ambientales entre las Aguas Continentales y las Marinas (p. 95–99). G. Capaldo, (Eds.), Mare Magnum, Buenos Aires, Argentina.

Billaud J.-P., Bruguière-Garde Y. (1992). "L'homme, l'eau, le Marais: quelques aspects d'une relation sociale dans une zone humide" dans *Economie Rurale*, **208–209,** 50–53.

Binet M.-E. and Carlevaro F., Michel P. (2012). Estimation of Residential Water Demand with Imprecise Price Perception, Working Paper, CREM, UMR, CNRS 6211.

Biswas A. (2005). "Integrated Water Resources Management: a reassessment". In: Integrated Water Resources Management in South and Southeast Asia. A. K. Biswas, O. Varis and C. Tortajada (eds.), Oxford University Press, New Delhi, India, pp. 325–341.

Biswas A. K. (2001). "World Water Forum: In Retrospect". *Water Policy*, **3**, 351–356.

Biswas A. K. (1993). "Management of international waters: problems and perspective". *International Journal of Water Resources Development*, **9**(2), 167–188.

Bjerström E. (2002). "Lack of awareness not only the failure of media". *Water Science and Technology*, **45**(8), 233–234.

Black M. and King J. (2009). The Atlas of Water. Second Edition: Mapping the World's Most Critical Resource. University of California Press, USA.

Blomqvist A. (2004). How can stakeholder participation improve European watershed management? The Water Framework Directive, watercourse groups and Swedish contributions to Baltic Sea eutrophication. *Water Policy*, **6**, 39–52.

Blot F. (2005). "Discours et pratiques autour du Développement Durable et des Ressources en Eau", Université Toulouse II - Le Mirail, Ecole Nationale de Formation Agronomique (ENFA), Ecole Nationale Supérieur, France.

Blouin Genest G. et Julien F., Paquerot S. (2012). L'eau en commun: de ressource naturelle à chose cosmopolitique. Presses de l'Université du Québec, Québec, Canada.

Boelens R. and Post Uiterweer N. C. (2013). "Hydraulic heroes: the ironies of utopian hydraulism and its politics of autonomy in the Guadalhorce Valley, Spain". *Journal of historical geography*, **41**, 44–58.

Bonnafous-Boucher, M. et Pesqueux Y. (2006). Décider avec les parties prenantes – Approche d'une nouvelle théorie de la société civile. La découverte, Paris, France.

Bord R. J. and O'Connor R. E. (1990). Risk communication, knowledge, and attitudes: Explaining reactions to a technology perceived as risky. *Risk Analysis*, **10**, 499–506.

Bougnoux D. (2002). Introduction aux sciences de la communication. La Découverte, Paris, France.

Bouleau L. et Guérin-Schneider G. (2011). Des tuyaux et des hommes: les réseaux d'eau en France. Éditions Quae, NSS dialogues, France.

Bourg D. (1997). Nature et technique. Essai sur l'idée du progrès. Patier, Paris, France.

Bourg D. (1996). L'homme-artifice. Le sens de la technique. Gallimard, Paris, France.

Bourg D. (dir.) (1993). La nature en politique ou l'enjeu philosophique de l'écologie. L'Harmattan, Paris, France.

Boutelet M., Larceneux A. et Barczak A. (2010). Gouvernance de l'eau. Intercommunalités et recomposition des territoires. Editions universitaires de Dijon, Dijon, France.

Boykoff M. (2011). Who Speaks for Climate? Making Sense of Media Reporting on Climate Change, Cambridge University Press, Cambridge, UK.

Bradshaw G. A. and Borchers J. G. (2000). "Uncertainty as information: narrowing the science-policy gap". *Conservation Ecology*, **4**(1), 7.

Brei V. A. and S. Böhm (2011)" Corporate Social Responsibility as Cultural Meaning Management: A Critique of the Marketing of 'Ethical' Bottled Water". *Business Ethics: A European Review*, **20**(3), 233–252.

Breton P. (2003). L'éloge de la parole. La Découverte, Paris, France.

Breton P. (1998). La parole manipulée. La Découverte, L Paris, France.

Briggs A. and P. Burke (2005). A Social History of the Media: From Gutenberg to the Internet. Ed. Polity Press, Malden, Ma, USA.

Bromberger C. et Pelen, J.-N. (1985). "Usages et Images de l'eau" dans *Le Monde Alpin et Rhodanien*, **4,** 7–14.

Brossard D. and Shanahan J. (2004). "Are Issue-Cycles Culturally Constructed? A Comparison of French and American Coverage of Global Climate Change". *Mass Communication & Society*, **7**(3), 359–377.

Brown R. and Farrelly M., Keath N. (2009). "Practitioner perceptions of social and institutional barriers to advancing a diverse water source approach in Australia". *International Journal of Water Resources Development*, **25**(1), 15–28.

Bruch C. (2004). "New tools for governing international watercourses". *Global Environmental Change*, **14**, Supplement, 15–23.

Bruchet A. and Burlingame G., Sklenar K. S., Suffet I. H., Whitfield F. B. (1999). Off-flavours in the Aquatic Environment 1997, Selected Proceedings of the 5th International Symposium on Off-flavours in the Aquatic Environment. *Water Science and Technology*, **40**(6), 1–300.

Bruns B. (2005). "Community-Based Principles for Negotiating Water Rights: Some Conjectures on Assumptions and Priorities". International Workshop on "African Water Laws: Plural Legislative Frameworks for Rural Water Management in Africa", 26–28 January 2005, Johannesburg, South Africa.

Bruvold W. H., Olson B. H. and Rigby M. (1981). "Public Policy for the Use of Reclaimed Water". Environmental Management, **5**(2), 95–107.

Bruvold W. H. and Ongerth H. J. (1974). "Public Use and Evaluation of Reclaimed Water". *Journal of the American Water Works Association*, **66**(5), 294–297.

Bruvold W. H. (1968). "Scales for the tasting of water". *Journal of Applied Psychology*, **32**, 45–253.

Bryant S. D. (2004). "Lead-contaminated drinking waters in the public schools of Philadelphia". *Journal of Toxicology, Clinic Toxicology*, **42**(3), 287–94.

Burgess J. (1990). "The Production and Consumption of Environmental Meanings in the Mass Media: A Research Agenda for the 1990s". *Transactions of the Institute for British Geography*, **15**, 139–161.

Burlingame G. A. and A. Tanjuncto (2001). Does Boiling Improve Water Quality and Taste? Opflow, March. 27, 3: 7, Proceedings AWWA Water Quality Technology Conference (WQTC), Colorado, USA.

Buurma H. (2001). "Public policy marketing: marketing exchange in the public sector". *European Journal of Marketing*, **35**(11/12), 1287–1300.

Buffet B. et Evrard R. (1950). L'eau potable à travers les ages. Editions Solédi, Liège, Belgique.

Callon M., Lascoumes P. et Barthe Y. (2001). Agir dans un monde incertain. Essai sur la démocratie technique. Seuil, Paris, France.

Camdessus M., Badre B., Cheret I. et Teniere-Buchot P.-F. (2004). Eau. Editions Robert Laffont, Paris, France.

Cans R. (2002). La ruée vers l'eau. Folio actuel, Paris, France.

Canter L. W., Nelson D. I. and Everett J. W. (1993/94). "Public perception of water-quality risks: Influencing factors and enhancement opportunities". *Journal of Environmental Systems*, **22**(2), 163– 187.

Caplan N. et al. (1975). The use of social science knowledge in policy decisions at the national level. The Institute for Social Research, Ann Arbor, MI, USA.

Care, J. W. (1975). "A cultural approach to communication". *Communication*, **2**, 1–22.

Carpe Diem Project (2009). "Water and Climate Change in the West, Polling and Media Analysis". Resource media, USA.

Carvalho A. and Burgess J., (2005). "Cultural Circuits of Climate Change in UK Broadsheet Newspapers, 1985 – 2003." *Risk Analysis*, **25**, 1457–69.

Casciarri B. et Van Aken M. (2013). Anthropologie et Eau(x), dans. *Journal des anthropologues*, 132–133.

Catellani A. (2011). "Environmentalists NGOs and the Construction of the Culprit: Semiotic Analysis". *Journal of Communication Management*, **15**(4), 280–297.

Catellani A. (2010). "La communication environnementale interne d'entreprise aujourd'hui: dissémination d'un nouveau 'grand récit'". dans *Communication et organisation*, **36**, 179–219.

Cazenave E. et Ulmann-Mauriat C. (1995). Presse, Radio et Télévision en France de 1631 à nos jours. Hachette, Paris, France.

CGEDD*, Monteils, I. and Rathouis, P. (2011). Accès à l'eau et à l'assainissement dans des conditions économiquement acceptables par tous, rapport de la mission sur la mise en œuvre de l'article 1 de la loi n 2006–1772 du 30 décembre sur l'eau et les milieux aquatiques, *Conseil Général de l'Environnement et du Développement Durable, France.

Chauhan S. S. (2003). "Environmental Education and Water Representation Among High School Students: A Case Study of Rajasthan (Jaipur Area)". *Journal of Human Ecology*, **14**(2), 113–117.

Chaussade, J.-L. et Pellay M. (2012). Les 100 mots de l'eau. Presses Universitaires de France, Paris, France.

Ching, L. and Yu D. J. H. (2009). Turning the tide: Informal Institutional Change in Water Reuse, Insitute of Water Policy, Working Paper Series, Serial n IWP/WP/No. 3/2009.

Ching L. (2010). "Eliminating Yuck: a simple exposition of media and social change in water reuse policies". *International Journal of Water Resources Development*, **26**(1), 111–124.

Chomsky N. (1957). Synthatic structures. Mouton & Co, The Hague, the Netherlands.

Chomsky N. (1969). American power and the New Mandarins. Pantheon, New York, USA.

Chomsky N. (1979). Language and responsibility. Pantheon, New York, USA.

Christen K. (2005). "Water Reuse: Getting Past the Yuck Factor". *Water Environment and Technology*, **17**(11), 11–15.

Clark G. A. and Espacio de Salud A. C. (1998). Dry sanitation in Morelos, Mexico, Paper presented at the Community Water Supply and Sanitation Conference. The World Bank, Washington D.C.

Commissariat Général au Développement Durable (2013). Guide du Dialogue avec les parties prenantes, Guides du Club DDEP, Ministère de l'Ecologie, du développement durable et l'énergie, France.

Conseil d'Etat (2011). Rapport public 2011: Consulter autrement, Participer effectivement, La documentation française, France.

Corral-Verdugo V., Bechtel R. B. and Fraijo-Sing B. (2003). "Environmental beliefs and water conservation: an empirical study". *Journal of Environmental Psychology*, **23**, 247–257.

Corral-Verdugo V. and Frias-Armenta M. (2006). "Personal Normative Beliefs, Antisocial Behavior, and Residential Water Conservation". *Environment and Behavior*, **38**(3), 406–421.

Corral-Verdugo V., Carrus G., Bonnes M., Moser G. and Sinha J. B. P. (2008). "Environmental Beliefs and Endorsement of Sustainable Development Principles in Water Conservation. Towards a New Human Interdependance Paradigm Scale". *Environment and Behavior*, **40**(5), 703–725.

Cortez-Lara A. A. (2011). "No longer strong social cohesion: Lessons from two transboundary water conflicts in the Mexicali Valley, Mexico". *Regional Integration and Social Cohesion, Berghahn Books*, **2**(2), New York/Oxford, USA/UK.

Cosgrove W. J. and F. R. Rijsberman (1998). "Creating a vision for water, life and the environment". *Water Policy*, **1**(1), 115–122.

CREDOC* (2011). Analyse du questionnaire: "Mise en œuvre de la solidarité pour l'eau", *Centre de Recherche pour l'Etude et l'Observation des Conditions de Vie, France.

Creighton J. L., Dunning C. M., Priscolli J. D. and Ayres D. B. (1998). Public Involvement and Dispute Resolution: A Reader on the Second Decade of Experience at the Institute for Water Resources. IWR Report 82-R-5. U.S Corps of Engineers, Alexandria, Virginia, p. 471.

Cronon W. (1992). "A Place for Stories: Nature, History, and Narrative". *Journal of American History*, **78,** 1347–76.

Crow D. A. (2012). Policy Diffusion and Innovation: Media and Experts in Colorado Recreational Water Rights. *Journal of Natural Resources Policy Research*, **4**(1), 27–41.

Crow D. A. (2010). "Local media and experts: Sources of environmental policy initiation?". *Policy Studies Journal*, **38**(1), 143–164.

Crow D. A. (2009). "Responsive Public Officials and Engaged Citizens: Myth or Reality? *A Case Study of Water Rights Policy in Colorado.* Public Organization Review, **9**(2), 119–138.

Damodaran N., Pratt J., Cromwell J., Lazo E. D., Raucher R., Herrick C., Rambo E., Deb A. and Snyder J. (2005). Customer Acceptance of Water Main Structural Reliability, Awwarf (Awwa Research Foundation), Denver, USA.

Davies S. R. (2008). "Constructing communication: Talking to scientists about talking to the public". *Science Communication*, **29**, 413–434.

Deeb R. A. (2009). Principles and Practices, Public Perception, and Message Effectiveness, Water Environment Research Foundation Report CEC2R08C, IWA Publishing, London, June 1.

Delorme D. E., Hagen S. C. and Stout I. J. (2003). Consumers' Perspectives on Water Issues: Directions for Educational Campaigns. *Journal of Environmental Education*, **34**(2), 28–35.

Demeritt D. (2006). "Science studies, climate change and the prospects for constructivist critique". *Economy and Society, 35(3)*, 453–479.

De Ouazzani S., Dugelay E. et Tutenuit C. (2012). L'entreprise et l'eau. Vers une gestion responsable, Etude Epe – Deloitte, Paris, France.

Dethman L., Dietemann A. and Lucas N. (2008). "A Better Way to Beautiful: Evidence that Education Changes Water Use Behavior", Presented at WaterSmart Innovations Conference 2008, Las Vegas, Nevada, USA.

Dethman L. and Thomley D. (2007). Residential Water Conservation Benchmarking Survey and Attribution/ Consumption Analysis. Seattle Public Utilities and Dethman and Tangora LLC. Seattle, Washington, USA.

De Young, R. (2000). "Expanding and Evaluating Motives for Environmentally Responsible Behavior". *Journal of Social Issues*, **56,** 509–526.

Dingfelder S. F. (2004). From toilet to tap: psychologists lend their expertise to overcoming the public's aversion to reclaimed water, APA Monitor **35** (8).

Dionet-Grivet S. (2011). Géopolitique de l'eau. Ellipses, Paris, France.

Dolnicar S. and Schafer A. (2007). "Desalination and Recycling: Australians raise health, environment and cost concerns". *Desalination and Water Reuse*, **16**(4), 10–15.

Dolnicar S. and Saunders C. (2005). "Marketing Recycled Water - Review of Past Studies and Research Agenda". In: S. J. Khan, M. H. Muston, A. I. Schaefer, (eds.), Integrated Concepts in Water Recycling, Conference Summary from the Organising Committee, Wollongong: Elsevier B. V., pp. 181–192.

Dooge J. (2004). Water Science and Technology: History and Future. Centre for Water Resources Research, University College, Dublin, Ireland.

Dorbeau-Falchier D., et al. (2007). "L'accès à l'eau potable dans les États fragiles". In: États et sociétés fragiles, J.-M. Châtaigner, H. Magro (sous la dir.), Karthala, Paris, France, pp. 407–418.

Douzou S. et Wilson K. G. (2010). Une Histoire des médias de Communication. Presses de l'Université du Québec, Québec, Canada.

Dubosc A. (2001). La participation du public dans les processus de décision liés à l'eau, Agence de l'eau Seine, Normandie, France.

Duchêne A. (2004). "Construction institutionnelle des discours. Idéologies et pratiques dans une organisation supranationale", Travaux neuchâtelois de linguistique, **40**, 93–115.

Ducroux A.-M. (2002). Les nouveaux utopistes du développement durable. Les éditions Autrement, Paris, France.

Duda M. D., De Michele P., Zurawski C., Jones M., Yoder J. E., Testerman W., Lanier A., Bissell S. J., Wang P. and Herrick J. (2003). Understanding The Georgia Public's Perception Of Water Issues And The Motivational Messages To Which They Will Respond, Responsive Management National Office, USA.

Dungumaro E. W. and Madulu N. F. (2003). "Public Participation in Integrated Water Resources Management: The Case of Tanzania". *Physics and Chemistry of the Earth*, **28,** 1009–1014.

Dunwoody S. (1999). "Scientists, Journalists, and the Meaning of Uncertainty". In: S. M. Friedman, S. Dunwoody and C. L. Rogers. Communicating Uncertainty: Media Coverage of New and Controversial Science, Mahwah, New Jersey, USA and Lawrence Erlbaum Associates, Inc. Publishers, London, UK, pp. 59–80.

Durand G. (1995). "Le symbolisme des eaux". In: Encyclopædia Universalis, Paris, **7**, 818.

Earle A. and Bazilli S. (2013). "A gendered critique of transboundary water management". *Feminist review*, **103**(1), 99–119.

El Ghassem O. A. (2001). Eléments pour une symbolique maure: de la dune au puits. L'Harmattan, Paris, France.

Espeland W. N. (1998). The struggle for water: politics, rationality, and identity in the American Southwest, University of Chicago Press, USA.

Fagan B. (2011). Elixir: A History of Water and Humankind. Bloomsbury Press, New York, USA.

Falahee M. and MacRae A. (1995). "Consumer appraisal of drinking water: multidimensional scaling analysis". *Food Quality and Preference*, **6**, 327- 332.

Fawell J. K. and Miller D. G. (1992). Drinking water quality and the consumer. *Journal of the Institution of Water and Environmental Management*, **6**, 726–732.

Fisher C. and Young J. (2011). Water for the Future Campaign, Final Evaluation Report, prepared for Department of Sustainability, Environment, Water, Populations and Communities.

Fischer F. (1993). "Citizen participation and the democratization of policy expertise: From theoretical inquiry to practical cases". *Policy Sciences*, **26**, 165–187.

Fischhoff B., Slovic P. and Lichtenstein S. (1978). "How Safe is Safe Enough? A Psychometric Study of Attitudes Towards Technological Risks and Benefits". *Policy Sciences*, **9**, 127–152.

Fischer G. W., Morgan M. G., Fischhoff B., Nair I. and Lave L. B. (1991). "What risks are people concerned about?" *Risk Analysis*, **11**, 303–31.

Fishman C. (2012). The Big Thirst: The Secret Life and Turbulent Future of Water. Free Press, USA.

Fitzsimmons J., Fong J., Johnson M. and Theel S. (2012). Media Avoid Climate Context In Wildfire Coverage, Media matters for America, USA.

Foddy M. and Smithson M. (1999). "Solving Natural Resource Dilemmas Through Structural Change: The Social Psychology of Metering Water Use". In: *Resolving Social Dilemmas Dynamic Structural and Intergroup Aspects*, M. Van-Vugt, (ed.), Psychology Press/Taylor & Francis, Philadelphia, USA, pp. 121–133.

Friedman T. (2008). Hot, flat and Crowded, Why We Need a Green Revolution - And How it Can Renew America, Farrar, Straus & Giroux, New York, USA.

Frohlich C. J. (2012). Security and discourse: the Israeli-Palestinian water conflict. *Conflict, Security & Development*, **12**(2), 123–148.

Funk A. (2006). "The Potential of Water Saving and Water Capturing Innovations: A Case Study of Albuquerque Single Family Homes." Professional Project Report Submitted in Partial Fulfillment of the Requirements for the Degree of Master of Water Resources.

Funk A. (2007). "Suggestions for Urban Water Conservation Planning: A New Mexico Perspective". *Bulletin of Science, Technology, and Society*, **27**(2), 170–181.

Galland F. (2011). Eau & conflictualités, Paris, France. In: Urban Ethnohydrology: Cultural Knowledge of Water Quality and Water Management in a Desert City, in *E&S HOME,* M. Choiseul.Gartin, B. Crona, A. Wutich and P. Westerhoff, (eds.), Vol. 15, N . 4, Article 36, 2010.

Gramsci A. (1971). Selections from the prison notebooks of Antonio Gramsci. International Publishers, New York: France.

Gaudin S. (2006). Effect of price information on residential water demand. *Applied Economy*, **38**(4), 383–93.

Gavin N. T., Leonard-Milsom L. and Montgomery J. (2011). "Climate Change, Flooding and the Media in Britain". *Public Understanding of Science*, **20**, 422–38.

GBSM (2011). Colorado's Water Future: A Communications Roadmap for Enhancing the Value of Water, for the Colorado Water Conservation Board, USA.

Geller E. S., Winett R. and Everett P. (1982). Preserving the Environment: New Strategies for Behavioral Change, Pergamon Press, New York.

Geller E. S., Erickson J. B. and Buttram B. A. (1983). "Attempts to Promote Residential Water Conservation With Educational, Behavioral, and Engineering Strategies". *Population and Environment*, **6**, 96–112.

Getzler J. (2006). A History of Water Rights at Common Law, Oxford Scholarship online, University Press School, UK.

Gilbert C. (2003). "La fabrique des risques". *Cahiers Internationaux de Sociologie*, **CXIV**, 55–72.

Gilbertson M., Hurlimann A. and Dolnicar S. (2011). "Does water context influence behavior and attitudes to water conservation?". *Australasian Journal of Environmental Management*, **18**(1), 47–60.

Giles B. (2013). Top Academics and Institutions in Water Research 2013, State of the Market Report. LuxResearch, Luxemburg.

Gleick H. P. (1993). "Water and conflict: Fresh water resources and international security". *Journal of International Security*, **18**.

Global Environmental Change Programme (1999). Designing 'interactive environmental research for wider social relevance, Special Briefing no 4, May 1999, University of Sussex, UK.

Godlewski A. (2010). "Damming" the Peace Process: Water Politics and its Impact on the Israeli-Palestinian Conflict. *Journal of Muslim minority affairs*, **30**(2), 153–166.

Goffman E. (1967). Interaction Ritual: Essays on Face-to-Face Behavior. Anchor Books, USA.

Goffman E. (1963). Behavior in Public Places: Notes on the Social Organization of Gatherings. The Free Press, USA.

Golubovska-Onisimova A., Flenley C. and Panina Y. (1999). Drinking water in Ukraine: communication and empowerment for local and international action, MAMA-86 and UNED, UK.

Gonzáleza A., Gilmera A., Foleyb R., Sweeneyc J. and Fryd J. (2008). "Technology-aided participative methods in environmental assessment: An international perspective". *Computers, Environment and Urban Systems*, **32**(4), 303–316.

Government of Orissa (2011). Orissa Urban Water and Sanitation Communication Strategy (Including Solid Waste Management), Government of Orissa, Housing and Urban Development Department, India.

Gramsci A. (1971). Selections from the prison notebooks of Antonio Gramsci. International Publishers, New York, France.

Granier F., Uhalde M. and Salles D. (2008). Agir en sociologie. Comprendre, débattre, concevoir, accompagner. Presses Universitaires de France, Paris, France.

Griffin R. J. and Dunwoody S. (2000). "The relation of communication to risk judgment and preventive behavior related to lead in tap water". *Health Communication*, **12**(1), 81–107.

Griskevivius V., Tybur J. M. and Van den Bergh B. (2010). Going green to be seen: Status, reputation, and conspicuous conservation. *Journal of Personality and Social Psychology*, **98**(3), 392–404.

Griswold B. and Warnow Z. (2009). "Communicating Climate Change and Water Linkages in the West", Resource Media, Seattle, USA.

Guillém F. (1975). Tribunal de las Aguas de Valencia y su proceso (Oralidad, concentración, rapidez, economía, www.tribunaldelasaguas.com).

Hamilton L. C. (1985). Self-reported and actual savings in a water conservation campaign. *Environment and Behavior*, **17**, 315–326.

Hansen M. T. (1999). "The search-transfer problem: The role of weak ties in sharing knowledge across organization subunits". *Administrative Science Quarterly*, **44**(1), 88–111.

Happs J. (1986). "Constructing an understanding of water quality: public perceptions and attitudes concerning three different water bodies". *Research in Science Education*, **16**, 208–215.

Harribey J.-M. (1998). Le Développement durable. Economica, Paris, France.

Harris D. H. (1977). "The Human Dimensions of Water-Resources Planning". *Human Factors*, **19**(3), 241–251.

Hartley T. W. (2006). "Public Perception and Participation in Water Reuse". *Desalinisation*, **187**(1–3), 115–116.

Hartley T. W. (2005). Public perception and participation in water reuse, Department of Resource Economics and Development, University of New Hampshire, USA.

Hartley T. W. (2003). "Water Reuse: Understanding Public Perception and Participation", Water Environment Research Foundation, Alexandria, VA, USA.

Hartley T. W. (2001). "Public Perception and Participation". *Water Reuse: Literature Summary*. Resolve Inc, Washington, D.C., USA.

Herbst S., Benedikter S., Koester U., Phan N., Berger C., Rechenburg A., Kistemann T. (2009). "Perceptions of water, sanitation and health: a case study from the Mekong Delta, Vietnam". *Water Science Technology*, **60**(3), 699–707.

Heinrich M. (2008). Water Use in *Auckland Households: Auckland Water Use Study* (AWUS) Final Report. BRANZ Limited, Judgeford, New Zealand.

Henderson-Sellers A. (1998). "Climate Whispers: Media Communication about Climate Change". *Climatic Change*, **40**, 421–456.

Heymann E., Lizio D. and Siehlow M. (2010). World water markets. Frankfurt, Deutsche Bank Research, Germany.

Hidiroglu P. (1994). L'eau divine et sa symbolique. Albin Michel, Paris, France.

Histace C. and Ravot N. (2012). "Faire connaître et animer la consultation du public". In: Comment communiquer, sensibiliser sur Sdage, Les rencontres de l'eau en Loire-Bretagne, Tours, France.

Hobson W. L., Knochel M. L., Byington C. L., Young P. C., Hoff C. J. and Buchi K. F. (2007). "Bottled, filtered, and tap water use in Latino and non-Latino children". *Arch Pediatr Adolesc Med*, **161**(5), 457–61.

Hoffman J. R. (2006). "Do We Have A Water Problem? The Use of Social Marketing as a Problem Solver". *Journal of the American Water Works Association*, August 2006, 34–36.

Holmes J. and Savgard J. (2008). Dissemination and implementation of environmental research, Swedish Environmental Protection Agency Report 5681, Bromma, Sweden.

House M. A. (1996). "Public perception and water quality management". *Water Science and Technology*, **34** (12), 25–32.

Howe C. (2012). Cultural Perspectives on the Rapidly Evolving Language of Water, SG: Public and Customer Communications, World Water Congress, Busan, South Korea.

Howe C. W. and Linaweaver F. P. (1967). The impact of price on residential water demand and its relation to system design and price structure. *Water Resource Research*, **3**(1), 13–32.

Hurd R. E. (2003). Consumer Attitude Survey on Water Quality Issues, AWWA (American Water Works Association), Research Foundation, Denver, CO, USA.

Hurlimann A. (2007). "Time for a water re-'vision'. Australasian". *Journal of Environmental Management*, **14**, 14–21.

Hurlimann A. (2007). "Urban versus rural – how public attitudes to recycled water differ in these contexts." Proceedings from the 6th International Water Association Specialty Conference on Wastewater Reclamation & Reuse for Sustainability, Antwerp Belgium, 9–12 October. CDROM.

International Human Rights Law Clinic (2013). The Human Right to Water Bill in California an implementation framework for state agencies, University of California, School of Law, Berkeley, USA.

Islam S. and Susskind L. (2013). Water diplomacy. A negociated approach to managing complex water networks, RFF Press, New York, USA.

Jackowska J. (2011). De l'eau et des idéaux. Les formes d'engagement des organisations de la société civile à l'ère de la gouvernance de l'eau au Sahel. Pistes de réflexion dans le domaine de l'accès à l'eau potable, sous la direction de Michel Korinman, Thèse de Doctorat de Géographie, Paris, France.

Jagerskog A. (2003). "The power of the "sanctioned discourse" - a crucial factor in determining water policy". *Water science and technology*, **47**, part 6, 161–166.

Jalenques-Vigouroux B. (2007). "Quand les entreprises communiquent sur le développement durable: un choix libre ou contraint ?" *dans Quaderni*, **64**, 97–104.

Jambes J.-P. (2000). "Territoire en questions: doutes et réponses de la géographie". In: Territoires sous influence. L'Harmattan, Paris, France.

Jeffery J. (2001). Understanding public receptivity issues regarding "in-house" water recycling. Results from a UK survey, unpublished manuscript, Cranfield University, UK.

Jeffries L. (2003). "Not a drop to drink: Emerging meanings in local newspaper reporting of the 1995 water crisis in Yorkshire". *An interdisciplinary Journal for the study of discourse*, **23**, part 4, 513–538.

Jesperson K. (2005). There's An Easier Way to Save Water. Accentuate The Positive: How Social Marketing Can Make A Difference, On Tap, Fall 2005.

Johnson N., Munk Ravnborg H., Westerman O. and Probst K. (2001). "User participation in watershed management and research". *Water Policy*, **3**, 507–520.

Johnson R. J. and Scicchitano M. J. (2000). "Uncertainty, risk, trust and information: public perceptions of environmental issues and willingness to take action". *Policy Studies Journal*, **28**(3), 633– 647.

Jones A. Q., Dewey C. E., Dore K., Majowicz S. E., McEwen S. A., Waltner-Toews D., Henson S. J. and Matthews E. (2005). "Public perception of drinking water from private water supplies: focus group analyses". *BMC Public Health*, **5**(129), 1–12.

Jozan R. (2012). Les débordements de la mer d'Aral: une sociologie de la guerre de l'eau. Paris, France: Presses universitaires de France.

Kantola S. J., Syme G. J. and Nesdale A. R. (1983). "The Effects of Appraised Severity and Efficacy in Promoting Water Conservation: An Informational Analysis". *Journal of Applied Social Psychology*, **13**(2), 164–182.

Kallis G. and Butler D. (2001). "The EU water framework directive: measures and implications". *Water Policy*, **3**, 125–142.

Karthikeyan G. B. and Surya Vardhan T. M. (2010). Consumer's Attitude and Perception towards Packaged Drinking Water (with special reference to Virudhunagar Town), Master Thesis, Faculty of the Virginia Polytechnic Institute and State University, USA.

Kasperson R. E., Renn O., Slovic P., Brown H. S., Emel J., Goble R., Kasperson J. X. and Ratick S. (1988). The Social Amplification of Risk A Conceptual Framework. *Risk Analysis*, **8**(2).

Kellner D. (2000). "Habermas, the public sphere and democracy: a critical intervention". In: Perspectives on Habermas, L. E. Hahn (ed.), Open Court, Chicago and La Salle, Illinois, USA, pp. 259–538.

Ketelaar A., Clarke T. and Armstrong D. (2012). Compliance and enforcement in water resource management project. Stakeholder analysis. Final Report, AK. Consultants.

Kirilenko A., et al. (2012). "Computer-assisted analysis of public discourse: a case study of the precautionary principle in the US and UK press". *Quality and Quantity*, **46**(2), 501–522.

Kiss A.-C. (1973). "La protection de l'environnement et les organisations européennes". *Annuaire français de droit international*, **19**(19), 895–921.

Kow J. K. (2001). "Television as a Tool of Environmental Adult Education: Limits and Possibilities". *Adult Education*, 30–41.

Kurz T., Donaghue N., Rapley M. and Walker I. (2005). "The Ways That People Talk About Natural Resources: Discursive Strategies as Barriers to Environmentally Sustainable Practices". *British Journal of Social Psychology*, **44,** 603–620.

Lagadec P. (1988). Défaillances technologiques et déstabilisations sociales. Seuil, Paris, France.

Lam S. P. (1999). Predicting Intentions to Conserve Water From the Theory of Planned Behavior, Perceived Moral Obligation, and Perceived Water Right. *Journal of Applied Social Psychology*, **29**(5), 1058–1071.

LaMoreaux P. E. and Powell W. J. (1996). "Springs and bottled waters". *Environmental Geology*, **27**(2), 75–76.

Lane J. (2000). "The 2nd World Water Forum at The Hague, March 2000". *Water Policy*, **2**(6), 465–467.

Langford I., Georgiou S., Bateman I., Day R. and Turner R. K. (2000). "Public Perceptions of Health Risks from Polluted Coastal Bathing Waters: A Mixed Methodological Analysis Using Cultural Theory". *Risk Analysis*, **20,** 691–704.

Larbi Bouguerra M. (2000). Sept propositions pour la gouvernance de l'eau, Fondation Charles-Léopold Mayer pour le progrès de l'homme, Paris-Lausanne, France-Suisse.

Larbi Bouguerra M. (2003). Les batailles de l'eau. Pour le bien de l'humanité. Enjeux Planètes, Paris, France.

Lascoumes P. et Le Gales P. (2009). Sociologie de l'action publique. Armand Colin, Paris, France.

Lasserre F. et Descroix L. (2011). Eaux et territoires: tensions, coopérations et géopolitique de l'eau, Thèse de doctorat de Géographie, avec la collaboration de Jean Burton et Anne Le Strat, Université Paris 4, France.

Lawton M., Birchfield D. et Wilson D. (2008). Slowing the Flow: A comprehensive demand management framework for reticulated water supply. Auckland, Beacon Pathway Limited, New Zeland.

Lefeuvre J.-C. et Pion N. (2009). L'eau douce en France: Histoire d'un long combat. In: Milan (ed.), Terre sauvage, France.

Lelièvre M. (2010). "Le développement durable, de la concertation à la codécision". *Responsabilité et environnement*, **60**, 12–14.

Le Net M. (1993). Communication publique: Pratique des campagnes d'information, Les Études de la documentation Française, France.

Levallois P., Grondin J. and Gingras S. (1999). Evaluation of consumer attitudes on taste and tap water alternatives in Quebec. *Water Science Technology*, **40,** 135–139.

Levallois P., Guevin N., Gingras S., Levesque B., Weber J.-P. and Letarte R. (1998). New patterns of drinking water consumption: results of a pilot study. *Science Total Environment*, **209**, 233–241.

Leveque C. and Van Der Leuuw S. (2003). Quelles natures voulons-nous? Pour une approche socio-écologique du champ de l'environnement. Elservier, Paris, France.

Lievrouw L. A. (1992). Communication, Representation, and Scientific Knowledge: A Conceptual Framework and Case Study Knowledge and Policy. *The International Journal of Knowledge Transfer and Utilization*, **5**(1).

Limbert M. (2001). "Senses of Water in an Omani Town". *Social Text 68*, **19**(3), 35–55.

Lohman L. C. and Milliken J. G. (1984). Public attitudes toward potable wastewater reuse: a longitudinal case study. Proc. Water Reuse Symposium III: Future of Water Reuse, **1**, 109–121.

Long J. (2001). Desecuritizing the water issue in *Singapore–Malaysia relations. Contemporary Southeast Asia*, **23**(3), 504–532.

Luan I. O. B. (2010). Singapore Water Management Policies and Practices. *International Journal of Water Resources Development*, **26**(1), 65–80.

Lytle L., Varnell S., Murray D., Story M., Perry C., Birnbaum A. and Kubik M. (2003). "Predicting adolescents' intake of fruits and vegetables". *Journal of Nutruition Education and Behavior,* **35**(4), 170–178.

Mac Luhan M. (1970). From Cliche to Archetype. The Viking Press, New York, USA.

Mac Luhan M. (1964). Understanding Media: The Extensions of Man. McGraw-Hill, New-York, USA.

Mac Luhan M. (1962). The Gutenberg Galaxy: The Making of Typographic Man, University of Toronto Press, Canada.

Maganda C. and Koff H. (2011). Water management, migration and governance, in Regional Integration and Social Cohesion, **2**(2), Berghahn Books, New York/Oxford.

Malissar A. (2002). Les Romains et l'eau. Les Belles Lettres, Paris, France.

Maneglier H. (1990). Histoire de l'eau: du mythe à la pollution. Editions François Bourin, Paris, France.

Mankad A. and Tapsuwan S. (2010). "Alternative water supply sources: A review of social and economic drivers of community acceptance and adoption". In: *Ozwater'10,* Brisbane, 8–10 March 2010.

Marcondes L., C. Cohen C., Chaperon S., Falquet J. J., et al. (2011). Eau et féminismes: petite histoire croisée de la domination des femmes et de la nature. La Dispute, Paris, France.

Margat J. et Andreassian V. (2010). Une ressource naturelle pas comme les autres, dans Dossier: L'eau, une ressource précieuse, Analyse financière, 36, juillet-septembre 2010.

Marks J., Martin B. and Zadoroznyj M. (2008). "How Australians order acceptance of recycled water: National baseline data". *Journal of Sociology*, **44**(1), 83–99.

Marks J. (2004). "Back to the Future: Reviewing the Findings on Acceptance of Reclaimed Water", Proceedings from Enviroment Conference, Sydney, 28 March – April 1, 2004, CDROM.

Marks J. (2004). "Advancing Community Acceptance of Reclaimed Water". *Water Journal of the Australian Water Association*, **31**(5), 46–51.

Marquez C., Izquierdo M. and Espinet M. (2006). "Multimodal science teachers discourse in modeling the water cycle". *Science education*, **90**(2), 202–226.

Marsilly (De), G. (1985). L'eau. Flammarion-Dominos, Paris, France.

Mathis M. L. (1999). "Using New Approaches to Environmental Decision-Making: An Application of Integrated Assessment Methods to Water Resource Issues in the Binational in the Lower Rio Grande Basin". *Policy Studies Review*, **16**(3&4), 139–167.

Masson D. (1985). L'eau, le feu, la lumière. Desclée De Browe, Paris, France.

Mattelard A. (2006, ed. 2001) Histoire de la société de l'information. La Découverte, Paris, France.

Mayer P. (2006). "Water Efficiency Benchmarking for New Single-Family Homes", Presented at the 2006 AWWA Water Sources Conference and Exposition, Albuquerque, New Mexico, USA.

Mayer P., DeOreo W. B., et al. (1999). Residential End Uses of Water. American Water Works Association, Colorado, USA.

McComas K. and Shanahan J. (1999). "Telling Stories About Global Climate Change: Measuring the Impact of Narratives on Issue Cycles". *Communication Research*, **26**(1), 30–57.

McCully P. (1996). Silenced Rivers: the ecology and politics and large dams. Zed Books Limited, New York, USA.

McDaniel M. (2001). Public Perception and Participation in Water Reuse. Hartley, Ed., RESOLVE Inc, Washington DC, USA.

McKinsey (2009). The global corporate water footprint. McKinsey Company, USA.

McManus P. A. (2000). "Beyond Kyoto? Media Representation of an Environmental Issue". *Australian Geographical Studies*, **38**(3), 306–319.

Means E., Chowdbury Z., Westerhoff G., Passantino L. and Ruettan J. (2009). Communicating the Value of Water: An Introductory Guide for Utilities, Awwa Research Foundation and the US Environmental Protection Agency, Denver, USA.

Media Monitors. (2007). "Analysis of media coverage of water – Experts respond". In: *Water in Australia, Media Monitors, Research Report*, Australia.

Menegaki A. N., Mellon R. C., Vrentzou A., Koumakis G. and Tsagarakis P. K. (2009). "What's in a name: Framing treated wastewater as recycled water increases willingness to use and willingness to pay". *Journal of Economic Psychology*, **30**, 285–292.

Merad M. (2011). Processus d'aide à la décision en gestion des risques - De la conduite du processus d'expertise à sa gouvernance, Thèse d'Habilitation à Diriger les Recherches (HDR), Paris IX Dauphine, France.

Merad M., Dechy N., Guionnet D. et Marcel F. (2009). L'Organisation face aux défis du développement durable: Cas de la mise en place une démarche de Développement Durable dans un institut public d'expertise dans le domaine de l'environnement industriel et des risques, Article –Note INERIS.

Merrill Singer M. and Erickson E. (2010). A Companion to Medical Anthropology. Blackwell Publishers, USA.

Michel-Guillou E. (2011). "La construction sociale de la ressource en eau". dans *Pratiques psychologiques*, **17**(3), 219–236.

Mikaïl B. (2008). L'eau, source de menaces? (ed.), Dalloz-Sirey, Paris, France.

Milliken J. (1999). "The Study of Discourse in International Relations: A Critique of Research and Methods". *European Journal of International Relations*, **5**, 225–254.

Milstein T. (2009). Environmental Communication Theories. In: Encyclopedia of Communication Theory. S. Littlejohn and K. Foss (eds.), Thousand Oaks, Sage UK, pp. 344–349.

Ministère de l'Écologie, du Développement durable, des Transports et du Logement (2011), Rapport du Gouvernement au Parlement sur la mise en place d'une allocation de solidarité eau, en application de la loi n 2011–156 du 7 février 2011 relative aux solidarités dans les domaines de l'alimentation en eau et de l'assainissement, France.

Ministry for the Environment (2009). On Tap? Attitudes, behaviours, and perceptions of household water use – informing demand management, Ministry for the Environment, Wellington, Australia.

Mirosa O. and Harris L. M. (2012). "Human Right to Water: Contemporary Challenges and Contours of a Global Debate". *Antipode*, **44**(3), 932–949.

Mitchell D. and Hanemann M. (1993). Setting Urban Water Rates for Efficiency and Conservation: A Discussion of Issues, California Urban Water Conservation Council, Sacramento, California, USA.

Mobley J., Tatham E., Reinhardt K. and Tatham C. (2006). Strategic Communication Planning: A Guide for Water Utilities, American Water Works Association Research Foundation, Denver, Colorado, USA.

Mobley J., Tatham E., Reinhardt K. and Tatham C. (2005). Message Management: Effective Communications, American Water Works Association Research Foundation, Denver, Colorado, USA.

Mollinga P. P. (2010). "The Material Conditions of a Polarized Discourse: Clamours and Silences in Critical Analysis of Agricultural Water Use in India". *Journal of agrarian change*, **10**(3), 414–443.

Mollinga P. P. (2008). "Water, Politics and Development: Framing a Political Sociology of Water Resources Management". *Water Alternatives*, **1**(1), 7–23.

Moore S., Murphy M. and Watson R. (1994). A Longitudinal Study of Domestic Water Conservation Behavior. Population and Environment. *A Journal of Interdisciplinary Studies*, **16**(2), 175–189.

Morali D. (1997). Anthropologie de l'eau. Presses Universitaires de Nancy, Nancy, France.

Murphy E. A. (1993). Effectiveness of flushing on reducing lead and copper levels in school drinking water. *Environnment Health Perspective*, **101**(3), 240–1.

Murray S. A. (2005). "Working for Water's 'Alienbusters': Material and Metaphoric Campaigns Against Alien Invaders". *Critical Alerts*, **19**(1&2), 127–149.

Nauges C. and Van Den Berg C. (2009). Perception of Health Risk and Averting Behavior: An Analysis of Household Water Consumption in Southwest Sri Lanka. *TSEWP Working paper series*, 09–139.

O'Donnell C. and Rice R. E. (2012). A Communication Approach to Campus Bottled Water Campaigns. *Social Marketing Quaterly*, **18**(4), 255–273.

O'Donnell C. and Rice R. E. (2012). A Communication Approach to Campus Mutin G. (2011) L'eau dans le monde arabe: menaces, enjeux, conflits. 2e edition. Ellipses, Paris, France.

Okbazghi Y. (2008). Water resources and inter-riparian relations in the Nile basin: the search for an integrative discourse. State University of New York Press and Bristol: University Presses Marketing, USA.

O'Laughlin J., McGuire K. and Carlson J. (1998). Public opinion of water quality issues and funding options in Idaho (Report 17). Moscow, ID: University of Idaho, Idaho Forest, Wildlife, and Range Policy Analysis Group, USA.

Olausson U. (2010). "Towards a European identity? The News Media and the Case of Climate Change". *European Journal of Communication*, **25**, 138–52.

Olivera O. and Lewis T. (2004). ¡COCHABAMBA! Water war in Bolivia. South End Press, Cambridge, U.K.

Olli E., Grendstad G. and Wollebaek D. (2001). "Correlates of Environmental Behaviors: Bringing Back the Social Context". *Environment and Behavior*, **33**(2), 181–208.

Ollivier-Yaniv C. (2006). "La communication publique: communication d'intérêt général et exercice du pouvoir". dans Olivesi S. (2006), Sciences de l'information et de la communication. Objets, savoirs, discipline. Presses universitaires de Grenoble, pp. 97–112.

Orisakwe O. E. and Frazzoli C. (2011). Water supply in Niger Delta of Nigeria: from public protests to scientific discourse. Orish Ebere Orisakwe and Chiara. NY, Nova Science Publishers USA.

Orlove B. and Steven C. (2010). "Water Sustainability: Anthropological Approaches and Prospects". *Annual Review of Anthropology*, **39**, 401–415.

Orne-Gliemann M. (2011). Des représentations de la gestion locale de l'eau: étude des discours et représentations à l'œuvre dans la mise en place de la réforme institutionnelle de l'eau en Afrique du Sud au sein de petits périmètres irrigués, Thèse de Doctorat, Université Paul Valéry – Montpellier III, France.

Osborne D., Walker D. and Sparham L. (2010). Water for the Future Communication Campaign Research. In: *Research Report*, Woolcott research.

Owen A. J., Colbourne J. S., Clayton C. R. I. and Fife-Schaw C. (1999). A mental models approach to customer perception of drinking water supply and quality. *Journal of the Chartered Institution of Water and Environment Management*, **13**(4), 241–244.

Pan Z. and Shah D. V. (1991). "Multilevel analysis in mass communication research" *Communication Research*, **18**(2), 140–174.

Pangborn R. M. and Bertolero L. L. (1972). Influence of Temperature on Taste Intensity and Degree of Liking of Drinking Water, JAWWA, **64**, 511–515.

Park E., Scherer C. W. and Glynn C. J. (2001). Community involvement and risk perception at personal and societal levels. *Health, Risk & Society*, **3**(3), 281–292.

Parkin R., Embrey M. A. and Hunter P. R. (2003). Communicating water-related health risks: lessons learned and emerging issues. *Journal American Water Works Association*, **95**(7), 58–66.

Pastoureau M. (2000). Bleu. Histoire d'une couleur. Le Seuil, Paris, France.

Patel A., Bogart L. M., Uyeda K. E., Rabin A. and Schuster M. A. (2010). Perceptions about availability and adequacy of drinking water in a large California school district. *Preventing Chronic Disease*, **7**(2), A39.

Panjabi R. K. L. (1993). "From Stockholm to Rio. A comparison of the declaratory principles of international environemental law". *Denver Journal of International Law and Policy*, **21**(215), 236–245.

Pearce F. (2007). When the Rivers Run Dry: Water – The Defining Crisis of the Twenty-First century. Beacon Press, Boston, USA.

Perry J. and Vanderklein E. (1996). Cultural dimensions of water quality policy. In: Water Quality: Management of a Natural Resource, Blackwell Science, Chapter 25, Cambridge, MA, USA.

Pidgeon N., Hood C., Jones D., Turner B. and Gibson R. (1992). "Risk perception", In: *Risk: Analysis, Perception and Management*. The Royal Society, London, 89–134.

Po M. and Nancarrow B. E. (2004). Literature Review: Consumer Perceptions of the Use of Reclaimed Water for Horticultural Irrigation. CSIRO Land and Water, Perth, Australia.

Po M., Kaercher J. D. and Nancarrow B. E. (2003). Literature Review of Factors Influencing Public Perceptions of Water Reuse. CSIRO Land and Water, Perth, Australia.

Poirier J.-P. (2012). L'eau, objectif du millénaire ou l'incurie des Etats. Autre Temps Editions, Paris, France.

Potter E., Mackinnon A., McKenzie S. and McKay J. (2007). Fresh Water: New Perspectives on Water in Australia. Melbourne University Press, Carlton, Australia.

Prescott J. (1998). Comparisons of taste perceptions and preferences of Japanese and Australian consumers. *Food Quality and Preference*, **9**(6), 393–402.

Priscolli J. D. (1998). "Public involvement, conflict management, and dispute resolution in water resources and environmental decision making". In: Public Involvement and Dispute Resolution: A Reader on the Second Decade of Experience at the Institute for Water Resources, J. L. Creighton, C. Dunning, J. D. Priscolli, D. B. Ayres, (eds.). IWR Report 82-R-5. U.S. Corps of Engineers, Alexandria, Virginia, USA.

Propp V. (1970). Morphologie du conte. Points Essais. Seuil, Paris, France.

Prud'Homme A. (2011). The Ripple Effect: The Fate of Fresh water in the Twenty-First Century. Scribner, New York, USA.

Quevauviller P. (2010). Water Systems Science and Policy Interfacing. RSC Publishing, Cambridge, UK.

Rahaman M. & Varis O. (2005). Integrated water resources management: evolution, prospects and future challenges. *Sustainability: Science, Practice, & Policy*, **1**(1): 15–21.

Rahaman M. and Varis O. (2003). The ethics of water: some realities and future challenges. CD Proceedings of 2nd Dubrovnik Conference on Sustainable Development of Energy, Water and Environment Systems, 15–20 June 2003, Dubrovnik, Croatia.

Raj S. D. (2005). Bottled Water: How Safe Is It? *Water Environment Research*, **77**(7), 3013–3018.

Recycled Water Task Force (2003). White paper of the public information, education and outreach workgroup on better public involvement in the recycled water decision process. Retrieved June 10, 2003, from the State of California Department of Water resources, USA.

Renn O., Webber T. and Johnson B. B. (1991). "Public Participation in Hazard Managment: The Use of Citizen Panels in the U.S." *Risk - Issues in Health and Safety*, **197**, 197–226.

Reymond P. (1958). L'eau, sa vie, et sa signification dans l'Ancien Testament. E. J. Brill, Leide, Belgique.

Risk solutions and DEFRA. (2009). Good Practice for Communicating about Drinking Water Quality. A report for The Drinking Water Inspectorate. London, UK.

Robinson K. G. Robinson C. H. and Hawkins S. A. (2005). "Assessment of public perception regarding wastewater reuse". *Water Science and Technology: Water Supply*, **5**(1), 59–65.

Rodda J. C. (2004). The basis of civilization, water science? In: *International Association of Hydrological Sciences*, IAHS Press, Center for Ecology and Hydrology, Wallingford, UK.

Roca E. and Villares M. (2008). Public perceptions for evaluating beach quality in urban and semi-natural environments. *Ocean & Coastal Management*, **51**, 314–329.

Rodriguez S. (2006). Acequia: Water Sharing, Sanctity, and Place. School for Advanced Research Press, Santa Fe, New Mexico, USA.

Rouse H., Ince S. (1957). History of hydraulics. Iowa Institute of Hydralic Research, Iowa, USA.

Royal Society (2006). Survey of factors affecting science communication by scientists and engineers. The Royal Society, London, UK.

Rudhart J. (1971). Le thème de l'eau primordiale dans la Mythologie Grecque. Société Suisse des Sciences Humaines, Berne, Suisse.

Russell A. S. (2008). An empirical investigation into the role of human rights in development discourse at the international level: a case study of the right to water. University of Oxford, UK.

Russell S. and Lux C. (2009). "Getting over yuck: Moving from psychological to cultural and sociotechnical analyses of responses to water recycling". *Water Policy*, **11**(1), 21–35.

Saito L., Berry K. D., Kauneckis D., and Berry A. K. (2011). "Understanding perceptions of successful cooperation on water quality issues: A comparison across six western U.S. interstate watersheds". In: Regional Integration and Social Cohesion, Volume 2, Issue 2, Berghahn Books, New York/Oxford.

Salles D. (2006). Les Défis de l'environnement. Démocratie et efficacité, Syllepse, Paris, France.

Sandman P. (2000). "Media Campaigns". In: Environmental Education & Communication for a Sustainable World, Chapter 11, B. A. Day and M. C. Monroe (eds.), Academy for Educational Development, NY, USA.

Sandman, P. (1974). "Mass environmental education: Can the media do the job". In: Environmental Education: Strategies Toward a More Liveable Future, J. A. Swan, et al. (eds.), John Wiley & Sons, New York, USA.

Salvetti M. and Winter C. (2012). Panorama des services publics d'eau et d'assainissement collectif en France. ONEMA, Paris, France.

Schaap W. and Van Steenbergen F. (2001). Ideas for Water Awareness Campaigns. Global Water Partnership, Stockholm, Sweden.

Schmid A., Thompson J. and Bengston D. (2007). "The Public Discourse About Land Use and Water Quality: Themes in Newspapers in the Upper Mississippi River Basin", *Applied environmental education and communication*, **6**(2), 187–196.

Schmidt C. W. (2008). The Yuck Factor: When Disgust meets Discovery. *Environmental Heath Perspectives*, **16**(12), A525-A527.

Sessa C. (2012). Sustainable Water Ecosystems Management in Europe. Bridging the Knowledge of Citizens, Scientists and Policy Makers. IWA Publishing, London, UK.

Sérodes J. (2011). 101 questions pratiques sur l'eau. Distribution du Nouveau monde. Éd. Multimondes, Québec, Canada.

Shaxson L., Alex T. Bielak, et al. 2012. Expanding our understanding of K*(KT, KE, KTT, KMb, KB, KM, etc.) A concept paper emerging from the K* conference held in Hamilton, Ontario, Canada, April 2012. UNU-INWEH, Hamilton, ON. Canada. http://inweh.unu.edu/wp-content/uploads/2013/05/KStar_ConceptPaper_FINAL_Oct29_WEBsmaller.pdf Accessed 23–10-2013

Shea L., Montillaud-Joyel S. (2005). Communicating Sustainability. How to produce effective public campaigns, UNEP (United Nations Environment Programme). Futerra Sustainability Communications Ltd, London, UK.

Sherrod D. (1998). "Using a stages of Change Framework to Assess Water Conservation Behavior", Proceedings of the AWWA Annual Conference, Dallas, Texas, USA.

Shiva V. (2010, first edition: 1998). Staying Alive: Women, Ecology, and Development. South End Press, Cambridge Massachusetts, USA.

Shiva V. (2002). Water Wars, Privatization, Pollution, and Profit. South End Press, Cambridge Massachusetts, USA.

Shiva V. (1993). Ecofeminism. Fernwood Publications, Halifax, Nova Scotia, Canada.

Sigworth E. A. (1957). Control of Odor and Taste in Water Supplies. *Journal of American Water Works Association*, **12**, 1507–1521.

Silva T., Pape D., Szoc R. and Mayer P. (2010). Water Conservation: Customer Behavior and Effective Communications, Water Research Foundation. U.S. Environmental Protection Agency, USA.

Silva T. (2006). "Selling Behavior Change: Examining Social Marketing Approaches For Water Conservation and Beyond". Proceedings of the AWWA Water Sources Conference, Albuquerque, New Mexico, USA.

Singh A. K. (2011). "Coverage of Development New in Print Media (A content analysis)". In: *Bilingual Journal of Humanities & Social Sciences*, Vol. 2, Issue 1 & 2 (Joint Issue), 15 Jan – 15 July 2011.

Smith C. (1990). "Social Marketing Strategies to Sell Water Conservation Programs". Proceedings of the AWWA Conservation Conference, Phoenix, Arizona, USA.

Soerens T. S., Fite E. H. and Hipp J. (2003). "Water quality in the Illinois River: Conflict and cooperation between Oklahoma and Arkansas". In: Diffuse Pollution and Basin Management, M. Bruen (ed.), Proceedings of the 7th International Specialised IWA Conference, Dublin, Ireland, p. 9–14.

Sofoulis Z. (2005). "Big Water, Everyday Water: A Sociotechnical Perspective". *Continuum: Journal of Media and Cultural Studies*, **19**(4), 445–463.

Sonnett J. et al. (2006). Drought and declining reservoirs: Comparing mediadiscourse in Arizona and New Mexico, 2002–2004. *Global environmental change: human and policy dimensions*, **16**(1), 95–113.

Soroczan C., Bach C. and Gauley B. (2004). "Educational Material on Outdoor Water Conservation: Thompson S. C. and S. Stoutemyer (1991) "Water Use as a Commons Dilemma: The Effects of Education That Focuses on Long-Term Consequences and Individual Action". *Environment and Behavior*, **23**(3), 314–333.

Sovocool K. A., Morgan M. and Bennett D. (2006). "An In-Depth Investigation of Xeriscape as a Water Conservation Measure". In: *Journal of the American Water Works Association*, USA.

Spence N. and Walters D. (2012). "Is it Safe? Risk Perception and Drinking Water in a Vulnerable Population". *The International Indigenous Policy Journal*, **3**(3).

Sprague de Camp L. (1993). The Ancient Engineers. Barnes and Nobles, New York, USA.

Spurgeon S. L. (2009). "Miracles in the Desert: Literature, Water, and Public Discourse in the American West". *Interdisciplinary studies in literature and environment*, **16**(4), 743–759.

Stenekes N., Schaefer A. and Ashbolt N. (2001). "Community Involvement in Water Recycling - Issues and Needs". In: Recent Advances in Water Recycling Technologies Workshop Proceedings, Brisbane: Centre for Water and Waste Technology, Civil and Environmental Engineering, A. I. Schaefer and T. D. Waite, P. Sherman (eds.), The University of New South Wales, Queensland Government Environmental Protection Authority, pp. 113–124.

Stewart G. (2009). The Case for Demand Management in Council Water Supplies. Wellington, New Zealand.

Stocking H. and Leonard J.-P. (1990). "The Greening of the Media". *Columbia Journalism Review*, 37–44.

Suarez J. M. (2011). "Achieving equitable water use in the Nile Basin: time to refocus the discourse on collective human security?" *Review of African political economy*, **38**(129), 455–466.

Subedi M. and Aryal M. (2010). Public perception about drinking jar water and its bacteriological analysis. *Nepal Med Coll*, **12**(2), 110–114.

Swain A. (2001). "Water wars: fact or fiction?" *The Journal of Futures*, **33**.

Syme G., Bennett D. and Kantola S. J. (1982). Public participation in Australian water resources planning, Hydrology & Water Resources Symposium 1982, Institution of Engineers Australia, 1–13 May, 1982.

Swallow B. M., Johnson N. and Meinzen-Dick R. S. (2001). Working with people for watershed management. *Water Policy*, **3**, 449–455.

Taylor M., Zaidah S. and Omar M. (2012). "Final Report on Media Desktop Analysis, Public Action for Water, Energy and Environment Project. Prosperity, Livelihoods and Conserving Ecosystems", IQC Task, Order #5, Report n 19, USAID / Jordan.

Thorpe A. (2006). "City of Atlanta Partners With State of Georgia to Help Water Customers Save Water and Money". Presented at the AWWA Water Sources Conference and Exposition, Albuquerque, New Mexico, USA.

TNS Conversa. Unpublished. Project Oracle: Water Usage and Conservation Across the Auckland Region, prepared for Watercare Services Limited by TNS Conversa in 2008, Auckland, New Zealand.

Ton A., Chapman M., Sansbury P. and Craig J. C. (2008). "An analysis of Media coverage on the Prevention and Early Detection of CKD in Australia". *American Journal of Kidney Diseases*, **52**(1), 159–170.

Tortajada C. (2004). Institutions for IWRM in Latin America. In: Integrated Water Resources Management in South and Southeast Asia, A. K. Biswas, O. Varis and C. Tortajada, (eds.), Oxford University Press, New Delhi, India, pp. 303–324.

Tran K. C., Euan J. and Isla M. L. (2002). Public perception of development issues: impact of water pollution on a small coastal community. *Ocean & Coastal Management*, **45**, 405–420.

Treitler I. and Midgett D. (2007). It's about Water: Anthropological Perspectives on Water and Policy. *Human Organisation*, **66**(2), 140–149.

Trumbo C. and O'Keefe G. J. (2001). "Intention to Conserve Water: Environmental Values, Planned Behavior, and Information Effects: A Comparison of Three Communities Sharing a Watershed". *Society and Natural Resources*, **14**, 889–899.

Trumbo C. (1996). "Constructing Climate Change: Claims and Frames in U.S. News Coverage of an Environmental Issue". *Public Understanding of Science*, **5**, 269–283.

Tsagarakis K. P. and Georgantzis N. (2003). "The role of information on farmers' willingness to use recycled water for irrigation". *Water Science and Technology: Water Supply*, **3**(4), 105–113.

Turgeon, S., Rodriguez M. J., Thériault M. and Levallois P. (2004). Perception of drinking water in the Quebec City region (Canada): The influence of water quality and consumer location in the distribution system. *Journal of Environmental Management*, **70**(4), 363–373.

UMR Research. Unpublished. Researching Public Opinion on Water Supply for Greater Wellington: A qualitative and quantitative study. Prepared for Greater Wellington Regional Council by UMR Research in 2007.

United Nations Centre for Human Settlements/ Habitat. (2001). 'Water Education in African Cities: report of an expert group meeting, 30 April-2 May 2001'

UNESCO. History, Development and Management of Water Resources, Volume 1 and 2. Encyclopedia of Life Support Systems, EOLSS, Series of publications, Paris, France.

Vaes G., Willems P., Swartenbroekx P., Kramer K., de Lange W., Kober K., Saravanan V. S., McDonald G. T. and Mollinga P. P. (2009). Critical review of Integrated Water Resources Management: Moving beyond polarised discourse. *Natural resources forum*, **33**(1), 76–86.

Valkenburg P. M., Semetko H. A. and De Vreese C. (1999). The effect of news frames on readers' recall and thoughts. *Communication Research*, **26**, 550–568.

Van der Beker A. (2002). "Water-related education, training and technology transfer". In: Knowledge for Sustainable Development, Vol. II. Encyclopedia of Life Support Systems Publications, Oxford, UK.

Van Vugt M. (2001). "Community Identification Moderating the Impact of Financial Incentives in a Natural Social Dilemma: Water Conservation". *Personality and Social Psychology Bulletin*, **27**, 1440–1450.

Varady R. G. and Milich L. L. (1999). "Openness, sustainability, and public participation: new designs for transboundary river basins". *Journal of Environment and Development*, **8**(3), 258.

Venko K. and Zdravka K. (2009). Les Médias en Europe Centrale et Orientale après 1989. Le Monde Diplomatique, Editions Bulgare, Sofia, Bulgarie.

Véron E. (1997). Construire L'evenement - Les médias et 'accident de Three Mile Island. Broché, Paris, France.

Vigneron J. et Francisco L. (1996). La communication environnementale. Collection Economica, Poche environnement, Paris, France.

Voorhees C. C. W., Vick J. and Perkins D. D. (2007). "Came Hell and High Water: The Intersection of Hurricane Katrina, the News Media, Race and Poverty". *Journal of Community & Applied Social Psychology*, **17**, 415–429.

Voropaev G. V. (1986). Socio-Economic Aspects of Water Resources Development in the USSR, Technical document in Hydrology. Unesco, Paris, France.

Wade B. (2000). "Tapping Into Water Shortage Solution". *American City and County* **115**(8), 45–48, 54–58.

Wahlberg A. A. and Sjöberg L. (2000). "Risk perception and the media". *Journal of Risk Research*, **3**(1), 31–50.

Walter B. (1939). L'œuvre d'art à l'époque de sa reproductibilité technique, nouvelle trad. Lionel Duvoy de la 4 version de l'essai de 1939. Allia, Paris, France.

Ward L. A., Cain O. L., Mullally R. A., et al. (2009). Health beliefs about bottled water: a qualitative study. *BMC Public Health*, **9**, 196–205.

Water Conservation Strategy Working Group (1998). A water conservation strategy for British Columbia. BC Ministry of Environment Lands and Parks, Canada.

Water Efficient Durham (1998). Outdoor Water Conservation, Project June–September 1998. Whitby. Water Efficient Durham, Ontario, Canada.

WBCSD (2009). Water for business: Initiatives guiding sustainable water management in the private sector. Weimann G. (1994). The influentials. People who influence people. State University of New York Press, Albany.

Weingart P., Engels, A., et al. (2000). "Risks of Communication: Discourses on Climate Change in Science, Politics, and the Mass Media". *Public Understanding of Science*, **9**, 261–283.

Wells D. L. (2005). Last but not least Perception, **34**, 1291–1292.

Wells D. L. (2005). "Last but not least: The identification and perception of bottled water". *Perception*, **34**(10), 1291–1292.

Weston S. (2010). Public Perception of Water Quality in Halifax Harbour. Saint Mary's University, 134.

Whelton A. J. and Dietrich A. M. (2001). Water Temperature as it Affects the Perception of Drinking Water Odorants, Proceedings AWWA Water Quality Technology Conference, Nashville, TN, November 11–15, p. 1–15.

Whiteford L. M. and Vindrola Pardos C. (2011). The Medical Anthropology of Water. In: A Companion to Medical Anthropology, M. Merrill Singer and E. Erickson, (eds.), Blackwell Publishers, New York, USA.

Wilbur J. (2006). Survey Explores Utah Attitudes and Knowledge About Water. *The Utah Watershed Review*, **14**(3), 1–4.

Wilkins L. (2000). "Searching for Symbolic Mitigation: Media Coverage of Two Floods". D. J. Parker (ed.), *Floods*, vol. II, pp. 218–244.

Wilkins L. and Patterson P. (1991). Risky business: communicating issues of science, risk, and public policy, Greenwood Press, USA.

Wilkins L. and Patterson P. (1987). "Risk Analysis and the construction of news". *Journal Of Communication*, **37**(3), 80–92.

Wilkins L. and Patterson P. (1991). "Risky Business: Communicating issues of Science". In: *Risk, and Public Policy*, Greenwood Press, USA.

Winkin Y. (1996). Anthropologie de la communication : de la théorie au terrain. Éditions De Boeck Université, Bruxelles, Belgique.

Willems P. and de Lange W. (2007). Concept of technical support to science–policy interfacing with respect to the implementation of the European water framework directive. *Environ. Sci. Policy*, **10**, 464–473.

Williams J. (2000). "The Phenomenology of Global Warming: The Role of Proposed Solutions as Competitive Factors in the Public Arenas of Discourse". *Human Ecology Review*, **7**(2), 63–72.

Wright J. A., Yang H. H., Rivett U. and Gundry S. W. (2012). Public perception of drinking water safety in South Africa 2002–2009: a repeated cross-sectional study. *BMC Public Health*, Jul 27, **12**, 556.

Wolton D. (2009). Informer n'est pas communiquer, CNRS, Paris, France.

Wolton D. (2003). L'autre mondialisation. Flammarion, Paris, France.

Wolton D. (1997). Penser la communication. Flammarion, Paris, France.

Wolton D. (1990). Éloge du grand public. Une théorie critique de la télévision. Flammarion, Paris, France.

World Water Council, The Secretariat of the 3rd World Water Forum. (2003). Contributions from Hoffman & Hoffman (H&H) and Michel Aublanc, in Analysis of the 3rd World Water Forum, World Water Council, Marseille, France, 14–16.

Yanovitzky I. and Stryker J. (2001). Mass Media, Social Norms and Health Promotion Efforts: A Longitudinal Study of Media Effects on Youth Binge Drinking. *Communication Research*, **28**(2), 208–239.

Yasar A., Khan N. Y., Batool A., Tabinda B., Mehmood R. and Women A. I. (2011). Perception of Water Quality and its Impacts on Health in Gangapur, Pakistan, Asian Network for Scientific Information.

Yearley S. (1996). Campaigning and Critique: Public-Interest Groups and Environmental Change. In: Earthly Goods: Environmental Change and Social Justice, F. O. Hampson and J. Reppy, (eds.), Cornell University Press, Ithaca and London, pp. 198–220.

Young C. C., Suffet I. H., Crozes G. and Bruchet A. (1999). Development of a Standard Method – Analysis of Compounds Causing Tastes and Odors in Drinking Water. *Water Science Technology*, **40**(6), 279–285.

Young G., Dooge J. and Rodda C. (2004). Global Water Resource Issues. University Press, Cambridge, UK.

Zehr S. (1999). "Scientists' Representations of Uncertainty". In: Communicating Uncertainty: Media Coverage of New and Controversial Science, S. M. Friedman, S. Dunwood and C. L. Rogers (eds.), Lawrence Erlbaum Associates, New Jersey, USA.

Zelem M.-C. (2010). Politiques de maîtrise de la demande d'énergie et résistances au changement. Une approche socio-anthropologique, L'Harmattan, Paris, France.

Zemp H. (2010). "Natural Hazards: Changing Media Environments and the Efficient Use of ICT for Disaster Communication". In: Advanced ICTs for Disaster Management and Threat Detection, E. Asimakopoulou, N. Bessis, (eds.), Collaborative and Distributed Frameworks, pp. 46–64.

Zinnbauer D., and Dobson R. (2008). Rapport mondial sur la corruption 2008. La corruption dans le secteur de l'eau. Transparency International, Berlin, Allemagne.

Zwarteveen M. Z. (1997). "Water: From Basic Need to Commodity: A Discussion on Gender and Water Rights in the Context of Irrigation". *World Development*, **25**(8), 1335–1349.

Glossary

Water Glossary

Service Providers
This term includes utilities that operate water and sewerage networks, the vast majority of which are operated by national, regional or local government departments or by publicly owned companies. The term also includes small-scale water and sanitation services (such as wells, standpipes or public toilets) that entrepreneurs, civil society organisations or community organisations normally operate. It also includes government departments that provide information and assistance to small-scale providers and households on sanitation and hygiene (COHRE, AAAS, SDC and UN-HABITAT (2007) *Manual on the Right to Water and Sanitation, Geneva, Switzerland*).

Private water sector
Any private companies involves in the management of water supply and sanitation through a contract with a public entity in charge of water resources management at the local, regional or national scale. The private water sector comprises private water operators of different sizes and from many different countries. They served 10% of the world population in 2013.

Public water sector
Any public utility involves in maintaining and managing water supply and sanitation. They are owned by the state, local authorities, and collective cooperatives, etc. They provide water and sanitation services to 90% of the world population in 2013.

Water users
Water users include the three types of main traditional water uses: agricultural (irrigation, farming, etc.), industries (heating, cooling, solvent, raw materials, etc.), and domestic uses.

Citizens
A resident of a city or town, especially one entitled to vote and enjoy other privileges there. Citizens are individuals, they both hold rights and responsibilities towards water.

Citizens Customers
A citizen customer is a customer of water services entitle to vote, to address its local, regional and national elected representative. Citizen customers are clients of a service but are also responsible to get information on water cycle management and to bring messages on water resources towards policy makers.

Water militant, Alterglobalist and altermondialist
As part of a global movement for global justice, they are opponents to capitalism and globalization. In the water sector, they in favor of the management of water resources by public utilities and the public sector as opposed to the private sector.

Information and Communication Glossary

Awareness	See Chapter 1, Box 1.1, p. 5
Actors	Entities with social impacts. They take part of communication process as both senders and receivers. They are also elements conveying significations to communication processes.
Communication	See Chapter 1, Box 1.1, p. 6
Dispute	See Chapter 1 A social process of interactions between two different parties.
Education	See Chapter 1, Box 1.1, p. 7
Information	See Chapter 1, Box 1.1, p. 6
Knowledge	See Chapter 1, Box 1.1, p. 6
Mediation	See Chapter 1, Box 1.1, p. 7
Public sphere	See Chapter 1 The public sphere is defined as an area where individuals come together to discuss societal problems.
Receivers	Individuals, communities, target groups, entities receiving the information and content from senders. They are listeners, readers, and observers in the communication process. They interpret the communication and can be considered as the audience.
Senders	Individuals, communities, target groups, entities initiating a communication process, creating the message, information and content towards receivers. They are speakers, writers, someone who gestures, etc. They provide the communication.
Target Group	A specific group of people within the target communication plan and strategy. They can be formed of people and organisations with different criteria (gender, marital status, revenues for people; size, status, objectives for organisations).

Index of Figures

Index of Tables

Index of Boxes

Annex 1 Methodology

Interviews

Different experts and professionals contributed into setting first research, campaigns and publication on communication about water from different perspectives; practitioners involved in communication strategies for their organizations or for Media; and younger researchers and professionals who are working with a social science background on water discourses and perceptions or journalism.

Christophe Audouin (Lyonnaise des Eaux, France)
Gerald Ahabwe (National Water and Sewerage Corporation, Uganda)
Orishaba Adellah (Journalist, Uganda)
Vladimir Arana (SIE, Canada)
Carmen Aguilar (University of Wisconsin, USA)
Dionysis Assimacopoulos (NTUA, Greece)
Alejandro Aponte (Mayo Publicidad, Peru)
Walter Esteban Barrutia Feijoo (ANA, Peru)
Alexandre Brailowski (SUEZ Environnement, France)
Constance Covillard (Degrémont, France)
Philippe Clapé (Rhone - Mediterranean Area - Corsica Agency, France)
Henri-Michel Cauchie (Research public Center, Gabriel Lippmann, Luxemburg)
Abenaitwe Cliff (Journalist, Uganda)
Eduardo Chamochumbi (ANA, Peru)
Russell Cuhel (University of Wisconsin, USA)
Cesar Centurion (Mayo Publicidad, Peru)
Jean-François Donzier (INBO, France)
Denis Desille (pS-Eau, France)
Céline Dubreuil (pS-Eau, France)
Renaud Douci (Solidarités International, France)
Francisco Dumler (ANA, Peru)
Frida Delgado (Grupo RPP, Peru)
Jean-Pierre Fontaine (University of Clermont Ferrand, France)
Vanessa Filhol (Lyonnaise des Eaux, France)
Nicoletta Forlano (FAO, Italy)
Pascale Guiffant (SUEZ Environnement, France)
Paola Grossi (NET Engineering Spa, Italy)
Bouchra Ghiati (Associated director, Morocco)
Jeanne Grueau (Lawyer, France)

Zelalem Gefero (Journalist, Ethiopia)
Jorge Gironas (Pontificia Universidad Catolica de Chile, Chile)
Emiline Hassenforder (IRSTEA, France)
Jean-Hugues Hermant (Veolia Eau, France)
Guillaume Haar (SAUR, France)
Christian Hunter (Pontificia Universidad Catolica de Chile, USA)
Charlotte Jacquot (Caisse des dépôts, France)
Hugo Jara (ANA, Peru)
Violetta Kuzmickaite (Expert, Lituania)
Prasanta Kalita (University of Illinois at Urbana Champagne, USA)
Kostas Kalabokidis (University of the Aegean, Greece)
Karen Kun (Waterlution, Canada)
Boris Lesjean (Kompetenz Zentrum Wasser Berlin, Germany)
Kim Le Quang (Militant, Belgium)
MariaLabaque (National University of Cordoba, Argentina)
Diane Lamoureux (Universite de Laval, Canada)
Frédérique Martini (ONEMA, France)
Josefina Maestu (UN-Water, Spain)
Belladonah Muzavazi (WATENN Consulting/WASH, Zimbabwe),
Alex and Tyler Mifflin (TV Producer, Canada)
Beatriz Medina (Amphos 21, Spain)
Benjamin Noury (Oxyo Water, France)
Thierry Noel (Amane Water, France)
Anselme Nkinsi (Journalist, R.D. Congo)
Joseph Ngome (Journalist, Kenya)
Luke Olang (Kenyatta University, Kenya)
James Ohayo (UN-Habitat, Kenya)
Juan Ocola (ANA, Peru)
Emmanuel Poilane (France Libertés, France)
Gwenael Prié (Consultant, France)
Chloe Parker (WaterWiki, IWA, UK)
Enrique Playan (CSIC, Spain)
João Pêgo (Faculdade de Engenharia da Universidade do Porto, Portugal)
Santiago Reyna (National University of Cordoba, Argentina)
Erik Struyf (ANA, Peru)
Vanessa Vereau (Gerente de Proyectos en Energia, Ambiente y Desarrollo, Peru)
Erik Van den Elsen (Wageningenur, the Netherlands)
Thomas Van Waeyenberge (Aquafed, Belgium)
May Wu (Argone National Laboratory, USA)
Rifaat Abdel Wahaab, Holding Company for Water & Wastewater (HCWW, Egypt)
Nery Zapata (CSIC, Spain)

Global discourses and conferences

We studied the titles of UN resolutions starting 1946 until 2012. Resolutions were selected based on their title according to the following key words: water, law, environment, development, human/man/woman. The systematic references to previous statements to helped us to better identify relevant resolutions about water resources. A total of 813 resolutions were referenced on 13.158 UN resolutions from 1946 until 2012 based on the analysis of their titles. We analysed the content of the UN Resolutions including references to water resources (fresh and salty water) e.g. 35 resolutions on water, 109 on environment and 107 resolutions related to water resources e.g. climate change and desertification.

We proceed to a discourse analysis of the major conferences organized by the UN since the first Earth Summit in Stockholm in 1972 until the last Earth Summit held in 2012 in Rio de Janeiro. These conferences were selected from their title and areas of application, the selection was made in a similar manner to above, water conferences were selected. Several conferences mentioned in the statements and forming part of the global context of challenges of water management were also used. Words describing these conferences are: environment (nature, desertification, pollution, climate change), development (economic, sustainable, human, social) people (housing, quality of life, human dignity) and women (gender). A total of 66 conference statements and UN Resolutions were analyzed.

We studied the institutional brochures and websites of the organizations of the UN system by targeting the following periods: publications dated 2000 on the occasion of the Millennium Development) for the Decade 2005-2015 "Water of life "in the major conferences on the environment and water over the period from 1972 to 2012 and during 1981-1990 the Decade for drinking water infrastructure and sanitation with the main limit of accessing to these resources. Several sites have allowed us to have access to these resources in particular UNDCWS, the Documentation Centre of the United Nations Water and Sanitation (UN-Documentation Center on Water and Sanitation) and IRC Centre Research International (International Research Center). A total of 191 brochures on the period 1979-2013 and the websites comprising the site of the UN-Water, the sites of water years (2007 to 2013), and the site of the UNESCO Decade "Water of life "have been identified. In this approach, we report the research we have done also on water and women and the different research on water issues since 2004 which has led us to regularly consulted the "water" pages or the "gender" pages, "woman", "child", "green economy", "Goals" of the following sites: the WHO, UNEP, UNDP, UN, UNICEF UN Habitat, the World Bank, FAO, UNESCO and in particular its programs IHP – International Hydrological Programme (International Hydrological Programme), the site of the MDGs – Millennium Development Goals, the site WWAP – World program for the evaluation of water resources (World Water Assessment Programme).

Communication campaigns on water

We covered an analysis of communication campaigns on water. We focused on France, Europe, the USA, Canada and Australia.

Our study focused on the period of the 2000s with a few campaigns on 80-90 years. **A first sample of more than 165 campaigns was established for the period of 2000-2013.** We consider that the collection and referencing of water campaigns must be developped in the future. Selected campaigns were carried out by different types of the following actors: organizations from the United Nations, humanitarian organizations, national government organizations, private companies water management, public/municipal water companies the mineral water companies and initiatives of individuals. We conducted the selection based on three different criteria: the results of the survey conducted with water professionals, their presence on Internet, and their global impact.

We additionally collected logos of water and wastewater companies by country. We identified **500 logos in total from 80 countries**, brochures, participations to conferences and exhibition events as well as twitter where many water organisations have accounts.

We conducted a research of guide and books about communication in the water sector. **We selected 35 guides developed by major organisations**: UNICEF, the World Bank, WHO, UN Water, European funded projects, OCDE, GTZ, USAID, AFD and national government initiatives in the USA (EPA), the UK (DEFRA), Australia, and France (Basin agency, SAGE).

We completed an analysis of book titles published in the water based on a literature search from the word "water." We conducted a double research bibliography from Amazon (under books) site to observe the non-scientific books available to the general public and from the catalog of the François Mitterrand Library. The first provides access to 1,200 references (limit given by the site) and the second lists 1,412 records printed. Depending on the summary and the topic, we need to make a categorization of publications around redundant topics. These securities may be classified into the following categories: "Water War", "Water of life", "Water Science", "governance", "geopolitics," "the right to water", "philosophical, sociological and religious aspects"," citizen participation", "water markets", "scandal" and "waste".

We complemented our study with online library and information centers. Here is a selective and non-exhaustive list. (The two last references are in French language):

- Water Library – UN Water: www.unwaterlibrary.org
- Water datas & statistics (World Bank): www.water.worldblank.org
  www.worldbank.org/en/topic/waterresourcesmanagement
- UNESCO-IHE online resource (UNESCO): www.unesco-ihe.org
- WWAP (UNESCO): www.unesco.org/water/wwap
  See complete list of resources centers worldwide (Water centers)
- UNU-INWEH online resource (UN): http://inweh.unu.edu/resources/
- Aquastat (FAO): www.fao.org/nr/aquastat
- GWP Toolbox (Global Water Partnership): www.gwptoolbox.org
- Water Information System for Europe (WISE): http://water.europa.eu/
- INBO Information resources (International Network of Basin Organisations): www.inbo-news.org
- Water and Climate Bibliography (Pacific Institute): http://biblio.pacinst.org/biblio/
- The World's Water (Pacific Institute): http://worldwater.org/
- Growing Blue (Veolia): http://growingblue.com/
- WASH Resources (IRC): http://washresources.wordpress.com/
- Water Knowledge (AWWA): www.awwa.org
- US Geological Survey (USGS): www.usgs.gov/water
- Water Resources Research - Geophysical data (American Geophysical Union): http://publications.agu.org/digital-library/
- Circle of Blue (Journalist – News): www.circleofblue.org
- Global Water Intelligence (Market): www.globalwaterintel.com/
- Wisconsin's Water Library (USA): http://aqua.wisc.edu/waterlibrary/
- Water Home (EPA online resources): http://water.epa.gov/
- International Water Resources Association (IWRA): www.iwra.org
- IWA Water Wiki (IWA): www.iwawaterwiki.org
- International Water Center, online resources (IWC): www.watercentre.org
- Water Resources Management Programme (CEDARE): http://water.cedare.int/
- Asia-Pacific Water Forum (links towards many online resources): www.apwf.org/link
- Pedag'eau – pS-Eau: www.pseau.org/outils/biblio
- Prix de l'eau (France Libertés): www.prixdeleau.fr

(All accessed between June and October 2013)

Media and entertainment materials

Media analysis is excluded from the reference corpus. The media are nevertheless present throughout the study, particularly in France where many campaigns are oriented around scandals water quality (pesticides, cancer, legionella, etc.), management (corruption, opposition private services/public), wars and crises related to water (conflicts, droughts) .. For media, the study shall include the press, television news programs, radio shows and media information, computerized information, blogs, social networks and hybrid media. A study of media requires a corpus and research remains to be done. However, several media scandals have been a first place ranking for conferences and campaigns in a media context. We studied five controversies between 1998 and 2013 in France on the issue of water quality combining a first body 50 items from the press coverage. Other references come from literature and published scientific work (see Bibliography).

The analysis of cultural industries is also excluded from the reference corpus. Movies, TV shows, documentaries, music production, or video games, are influencing public debates and perceptions. They play an essential role in the production and circulation of social representations, codes and messages on the water. Studies of audiovisual media, radio and multimedia productions on water have not been performed for this book. References are based on existing research. We however, include the movies and documentaries selected by the international water and films event organised since 2006, several blogs listing documentaries on water related issues; and information ressources and library.
(See International "Water and Film", Catalogues of movies)

Water professionals and interviews

We also conducted a qualitative survey with professionals from the water sector investigating on their perceptions of main water challenges and campaigns. We collected for the first survey, 70 answers from Water Communication professionals from 23 countries. We realized different research to gather materials from the variety of water stakeholders to further analyze their discourses, messages and campaigns. This work is also based on a research published in 2012 including 170 interviews from 59 countries with water professionals and citizens related to water projects. We completed several local studies in Congo, Egypt, Ethiopia, Morocco, Kenya between 2008 and 2013 as part as on-going process and research activities on water perceptions and communication projects.

In our attempt to cover different campaigns and projects, our study intends to be representative of research conducted in French and English; it is as such, restricted. It covers primarily research from France, the European Commission, U.S.A., Canada, and Australia.

Annex 2 International Water Chronology

1921 Convention and Statute on the Regime of Navigable Waterways of International Concern, Barcelona, Spain

1923 Convention relating to the Development of Hydraulic Power affecting more than one State and Protocol of Signature, Geneva, Switzerland

1949 Recommendation to the international law commission to include the regime of territorial waters in its list of topics to be given priority, Res. 374 IV

1952 Right to exploit freely natural wealth and ressources, Res. 626 VII

1954 International technical conference on the conservation of the living resources of the sea, Res. 900 IX

1957 International conference of plenipotentiaries to examine the law of the sea, Res. 1105 XI

1958 1st United Nations Conference on the Law of the Sea, Geneva, Switzerland

1958 Question of initiating a study of the juridical regime of historic waters, including historic bays, Res. 1306 XIII

1958 Convening of a second UN conference on the law of the sea, Res. 1307 XIII

1959 Preliminary studies on the legal problems relating to the utilization and use of international rivers, Res. 1401 XIV

1959 Study of the juridical regime of historic waters, including historic bay, Res. 1453 XIV

1962 Conference on problems of developing countries, Cairo, Egypt

1966 Resources of the Sea, Res. 2172 - XXI

1969 Coordination of marine activities, Res. 2580

1970 The convention on the conservation of European wildlife and natural habitats, European Council

1970 Progressive development and codification of the rules of international law relating to international watercourses, Res. 2669 XXV

1970 Reservation exclusively for peaceful purposes of the Sea-bed and the Ocean floor, Res. 2750 XXV

1970 Permanent sovereignty over natural resources of developing countries and expansion of domestic sources of accumulation for economic development, Res. 2692 XXV

1972 The United Nations Conference on the Human Environment, 1st Earth Summit, Stockholm, Sweden

1977 The United Nations Water Conference, Mar del Plata, Argentina

1979 The United Nations Conference on Desertification, Nairobi, Kenya

1979	Follow-up to and implementation of the Mar del Plata action plan of the United Nations Water Conference, Res. A/RES/34/191
1980	Proclamation of the International Drinking Water Supply and Sanitation Decade, Res. A/RES/35/18
1980	Implementation of the Plan of Action to Combat Desertification, Res. A/RES/35/73
1980	Implementation in the Sudano-Sahelian Region of the Plan of Action to Combat Desertification, Res. A/RES/35/72
1981	Study on financing the Plan of Action to Combat Desertification, Res. A/RES/36/191
1981	United Nations Conference on New and Renewable Sources of Energy, Res. A/RES/36/193
1981	Draft World Charter for Nature
1982	The United Nations Conference on the Human Environment, 2nd Earth Summit, Nairobi, Kenya
1987	Countries stricken by desertification and drought in Africa, Res. A/RES/42/188
1987	Brundtland Report
1989	Large-scale pelagic driftnet fishing and its impact on the living marine resources of the world's oceans and seas, Res. A/RES/44/225
1990	Global Consultation on Safe Water and Sanitation for the 1990s, New Delhi, India
1990	Co-operation in fisheries in Africa, Res. A/RES/45/184
1991	Women, environment, population and Sustainable Development, Res. A/RES/46/167
1992	Protection of global climate for present and future generations of mankind, Res. A/RES/47/195
1992	Celebration of World Water Day
1992	The International Conference on Waterand the Environment, Dublin, Ireland
1992	The United Nations Conference on Environment and Development, 3rd Earth Summit, Rio De Janeiro, Brazil
1994	International Conference on Population and Development, Cairo, Egypt
1995	Convention on Biological Diversity, Res. A/RES/49/117
1995	Preventive action and intensification of the struggle against malaria in developing countries, particularly in Africa, Res. A/RES/50/128
1996	Convention on the law of the non-navigational uses of international watercourses
1997	World Water Forum in Marrakech, Morocco
1998	International Conference on Water and Sustainable Development, Paris, France
2000	World Water Forum in La Haye, the Netherlands
2000	Millenium Development Goals
2001	The International Conference on Freshwater, Bonn, Germany
2002	3rd Earth Summit, Johannesburg, South Africa
2003	World Water Forum in Kyoto, Japan
2005	UN Decade on "Water, source of life"
2006	World Water Forum in Mexico, Mexico
2008	Expo 2008 – International Conference "Water and Sustainable Development" in Zaragoza, Spain
2009	World Water Forum in Istanbul, Turkey
2009	Muscat First Ministerial Forum on Water in Muscat, Sultanate of Oman
2009	UN International Conference "Raising Awareness and Communicating the Urgency to Act" in Zaragoza, Spain
2010	Dushanbe Conference on the Implementation of the International Decade for Action "Water for Life"

2010 UN Sixty-fourth General Assembly Plenary 108th Meeting: Resolution Recognizing Access to Clean Water, Sanitation as Human Right

2010 UN Human Rights Council - Res. 64/292 the Human Right to Water and Sanitation

2012 World Water Forum in Marseille, France

2012 4th Earth Summit, Rio +20

Annex 3: List of guides to communication campaigns on water and related topics

This list is a selection of guides, toolkits or guidelines. It is non exhaustive. It proposed a variety of topics

Water Research to Market (2013) Guidelines to speed-up the transfer of water related outputs to better implement the Water Directives, EU funded project.

Water and Sanitation Program (2012) Introductory Guide to Sanitation Marketing: Developing a Communication Campaign, WSP/the World Bank.

World Bank (2012) Communication for Water Sector Reform. Obstacles and Opportunities, Washington, USA.

AFD, pS-Eau, PDM (2012) Elaborer une stratégie concertée pour l'eau et l'assainissement dans les villes secondaires africaines, Paris, France.

AWARE (2011) Bridging the Science-Policy Gap: Best Practives in Citizen's Participation, EU funded project.

Commonwealth of Australia (2011) Talking about water: words and images that enhance understanding, National Water Commission, Canberra, Australia.

DEFRA (2011) How to make a good environmental claim. Building trust and confidence in Environmental Communication, London, UK.

WaterReuse Research Foundation (2011) Talking about water. Vocabulary and Images that Support Informed Decisions about Water Recycling and Desalination, Alexandria, USA.

PSI (2011) Designing a communication campaign for enhanced PoU (Point of Use) water disinfectants, Dharavi, Mumbai, India.

WSP (2011) Introductory Guide to Sanitation Marketing, World Bank, Washington, USA.

American Water Works Association (2010) Water Conservation Communications Guide, USA.

Futerra (2010) Branding biodiversity. The new nature message, Futerra sustainability communications, UK.

Texas Water Development Board (2010) Developing a Water Conservation Public Awareness Program: A Guide for Utilities, Water IQ, USA.

USAID (2010) Public Action for Water, Energy and Environment Project. Propsperity, Livelihoods, and Conserving Ecosystems, USAID/Jordan.

Resource media (2009) Communicating Climate Change and Water Linkages in the West. Guidelines and Toolkit, Seattle, USA.

Risk solutions (2009) Good Practice for Communicating about Drinking Water Quality, A report for the Drinking Water Inspectorate, DEFRA, London, UK.

WERF (2009) Communication principles and practices, Public Perception, and Message Effictiveness. Related to Human and Environmental Health Effects with a Focus on Trace Organic Compounds, Water Environment Research Foundation, Alexandria, USA.

AWWA Research Foundation and EPA (2008) Communicating the Value of Water: An Introductory Guide for Water Utilities, Denver, USA.

Water and Sanitation Program – Latin America and the Caribbean (2008) Communication, water and sanitation in Latin America. The contribution of Communication for Development in water resource management and service implementation projects, WSP/The World Bank.

WERF (2008) Emergency Communications with Your Local Government and Community, Water Environment Research Foundation, Alexandria, USA.

SPLASH (2008) Guidance note on transferring research knowledge into action, German Development Institute/European funded project, Europe.

SAGE Mayenne (2008) Guide pour la sensibilisation aux économies d'eau, Laval, France.

DEFRA (2007) Sustainable events guide, London, UK.

USAID (2007) Best practices in social marketing safe water solutions for households water treatment: lessons learnt from population services international field programs, Washington, USA.

DEFRA (2006) Your guide to communicating climate change, London, UK.

GTZ (2006) Strategic Communication for Sustainable Development, a conceptual overview, Eschborn, Germany.

WHO (2005) Celebrating Water for Life. An advocacy guide. Geneva, Switzerland.

Futerra & UNEP (2005) Communicating Sustainability. How to produce effective public campaigns, United Nations Environment Programme, Paris, France.

WERF (2004) Public Perception of Biosolids Recycling: Developing Public Participation and Earning Trust, Water Environment Research Foundation, Alexandria, USA.

Global Water Partnership (2002) Ideas for Water Awareness Campaigns, Stockholm, Sweden.

WHO (2002) Healthy Villages. A guide for communities and community health workers, Geneva, Switzerland.

AFD, pS-Eau (2002) La prise en compte du genre dans les projets d'adduction d'eau potable en milieux ruraux et semi-urbain, guide méthodologique, Paris, France.

Agence de l'eau Rhône Méditerranée Corse (2000) Comment aborder la dimension sociale de l'eau, Direction de la planification et de la programmation, France.

OCDE (1999) Environmental Communication. Applying Communication Tools towards Sustainable Development, Paris, France.

UNICEF (1999) A Manual on Communication for Water Supply and Environmental Sanitation Programmes, New York, USA.

Annex 4: List of "water campaigns"

As stated in introduction and methodology, this study focused on France, Europe, the USA, Canada and Australia. It is not exhaustive. The selection was made based on interviews and research on Internet through different key works. It also includes campaigns from major programs and organisations from the water sector. For Africa and South America, several studies and websites were consulted. They offered a panel of different initiatives. For Asia, the overview was based on interviews with contributors and major events.

Websites and materials have been accessed between June and October 2013.

Short recent history of campaigns related to water, health, sanitation (1949-1999)

Before 1981

1949 UNICEF: "Thank you" - Postcard (They reached 1 million in 1952) - World
1950s UNICEF: "GOBI" campaigns (Growth, Oral rehydration, Breastfeeding, Immunization) - World
1955 UNICEF/WHO: "Campaign against malaria", leading child killer - World
1960/70s UNICEF/WHO: first actions and advocacy work on health, drinking water and water access - World
1972 Clean Water Action: "Clean Water Action" - Citizen/Door-to-door campaigns – Advocacy/USA
1975 UNICEF: "International Women's Year" - World
1975 WHO: "Guidelines for surveillance of Drinking Water quality" - World
1977 Eau & Rivières: "Campaign for River's protection' – Public rally followed by TV interview/Event/France

1981 – 1991 UN Decade

International Drinking Water Supply and Sanitation Decade/"Water and Sanitation for All"

1982 UNICEF: "Watsan" approach
1982 UNICEF: "Benefit soccer game" – 77,000 people/1 billion TV viewers
1981-2 World Bank: "VLOM: village-level operation and maintenance"
1981-2 UNICEF: Child Survival Revolution
1986 UNICEF: SWACH programme - India
1987 UNICEF: "Change for good" - Fundraising campaign/World
1989 UNICEF: "ORT" program (Oral Rehydration Treatment)
1992 UNICEF: "Internation Children's Day of Broadcasting" – Public/Radio - TV broadcasting/World

See Online IRC WASH Library, list of programs, case studies, analysis of the decade starting in 1974 - http://www.washdoc.info/page/53887 (accessed 15-06-2013)

Online UNICEF - WASH Resources and publications providing information on campaigns and programs from the 80s: http://www.unicef.org/wash/ (accessed 15-06-2013)

After 1992

1992 California Water Awareness: Campaigns - Teachers/Students/Members/Tutorials/USA
1993 Water Basin of Seine – Normandy: "Help water" – Teachers/Toolkit/France
1995 Clean Water Action: "Dirty Water Act" - Policy makers/Sitting – Rallies/USA
1996 Water Basin of Seine – Normandy: "Water classes" – Teachers/Class book/France
1998 International Rivers Network: "International Day of Action for Rivers"/Policy makers/Events: Public gathering/World
1998 Ricard: "Respect water" – Visual/France
1999 Evian: "The babies Ballet" - TV spot/World

See Ideas for Water Awareness Campaigns, Global Water Partnership (Wouter Schaap & Frank van Steenbergen)

Dated from 2002, this book provides a list of 67 campaigns from 33 different countries and 7 projects with an international outreach. The campaigns are mostly conducted at the end of 90s and early 2000. www.groundwatermanagement.org/documents/ideasonwaterawareness.pdf

Since 2000

See

- A communications Roadmap for Enhancing the Value of Water (Colorado Water Conservation Board) – List of 22 communication and water campaigns in the USA (2011) - http://cwcbweblink.state.co.us/weblink/
- Communication, water and sanitation in Latin America (WSP) - List of 21 communication campaigns in Latin America (2008) - www.wsp.org
- Communicating the Value of Water: An Introductory Guide for Water Utilities (EPA – AwwaRF) – Extensive analysis of communication messages and processes. This publication is based on a survey of utilities and work of experts. It provides the study of 7 detailed case studies in the USA.

1. 2000 Aquatox 2000: "International school network on water toxicity" - Students/Education/Canada
2. 2000 Circle of Blue: "Information about water" – Policy makers – Journalists – Scientists – Public/Internet resources – Newsletters/USA - World
3. 2000 Thames Water/New Millennium Experience Company: "The Millennium Dome Watercycle experiment" – Public/Exhibition/UK
4. 2000 The City of San Diego: "Think Blue" – Public – Kids – Residents – Businesses/Information campaigns/San Diego - USA - www.sandiego.gov/thinkblue
5. 2000 IRC: "Stream" – International coalition and Resource Center for water and sanitation access – Water professionals/Program – Resource centers/World
6. 2001 Suez: "Water for all" – Policy makers/Press article - Brochure/World
7. 2001 WSSCC: "The Big Issue" – Policy makers/Brochure – 11 visuals/World
8. 2002 WaterAid: "Water Matters" (120,000 petitions) – Policy Makers/Petition/UK
9. 2002 European Commission: "Water is life, Leaflet on the European Directive on water" – Policy Makers/Leaflet – Internet/Europe
10. 2002 Lyonnaise des Eaux: "Water Service" - Magazine for clients/Print/France

11.	2002	Lyonnaise des Eaux: "Halieutica's Trophees" - Fishermen – Policy makers/Press relations – Contest/France
12.	2003	Water Aid: "Flush out Poverty" – 80,000 postcards – Policy makers/Petition/UK
13.	2003	Lydec: "Clean Beaches in Casablanca" – Public/Event/Morocco
14.	2003	Ministry of Transport, Public Works and Water Management: "The Netherlands live with Water" – Public/TV – Radio - Comics/The Netherlands
15.	2004	SEDIF: "Tap Water" – Public/Visual/Paris - France
16.	2005	UN-Water/UN: "Water for Life" – Decade – Multi-public/Multi-supports/World a."Celebrating Water for Life" – Water professionals/Guide/World
17.	2005	WaterAid: "Make Poverty History" (20,000) Toilet paper – Policy makers– Petition/UK
18.	2005	ADEME: "Everybody needs water, don't waste" – Public/Visual - Leaflets/France
19.	2005	FP2E: "Thank you tap water" – Public/Visual – TV - Radio/France
20.	2005	Eau de Paris: "Distribution of water carafes" – Public/Event/Paris - France
21.	2005	Neau: "World Water for World citizens" – Public/Bottle Tap Water/The Netherlands
22.	2006	WaterAid: "Petition to the European Union for the change of the EU's water and sanitation policy" (20,000) – Policy makers/Petition/UK – Europe
23.	2006	Solidarités International: "Drinking Water" - Humanitarian emergency - Journalists/TV spot - Press release/France
24.	2006	SEDIF: Water deserves to be known" – Public – Residents/Visual - Radio/France
25.	2006	Besançon City: "La Bisontine" – Public/Product – Press Release/Besançon - France
26.	2006	EPA: "Watersense" – Public/Products/USA
27.	2007	World Water Day, Facing water scarcity – Public/Internet - Events/World
28.	2007	UNICEF: "Bad water kills more children than war" – Public – Donors/Visual/World
29.	2007	UNICEF Tap Project: "1$ for clean drinking water (Donation)" – Public – Donors – Business/Donation – Door to door /USA
30.	2007	Waterwise & Water Group Promotions: "Four minutes shower" (5,000 timers) - Public/ TV spot - Internet/UK
31.	2007	Texas Water Development Board: "Water IQ" - Research projet converted to awareness campaign - Public – Policy Makers /Guide – Internet/Texas - USA - www.wateriq.org
32.	2007	Association of California Agencies: "California's Water: A crisis we can't ignore" - Public/TV - Radio/USA
33.	2007	WHO: "A little water goes a long way" – Public/Products – Glasses & Leaflet/China
34.	2007	Solidarites International: "Water, first cause of death" – Journalist – Public/Press release/ France
35.	2007	WaterAid: "Knitted Petition" - 100.000 blue squares – Policy makers - Public/Petition/ UK – World
36.	2007	Companhia Athletica: "A quicker bath can save lives" – Public – Custiomers/Visual/ Brazil
37.	2007	Cristalline: "Who thinks tap ater has good taste don't drink it" - Public - Customers/ Visual/France
38.	2007	WEF: "Water Is Life, and Infrastructure Makes It Happen" - Water professionals – Policy makers – Public/Leaflet – Materials/USA
39.	2007	SmartWater: "The water with all the answers" - Public – Customers/Visual/USA - World
40.	2008	World Water Day: "International Year of Sanitation" – Mutli-public/Multi-supports/ World

41. 2008 Public Private Patnership for Handwashing (PPPHW): “Global Handwashing Day” – Children – Public/Events/70 countires - World
42. 2008 Solidarités International: “votregouttedeau.org” – Public – Policy makers – Donors/ Internet – Petition/France
43. 2008 Charity Water: “Jennifer Connelly walks in Central Park to get dirty water” – Public – Donors/TV spot/USA
44. 2008 Lydec: “My Earth” – Public (4,000)/Art exhibition/Morocco
45. 2008 WaterAid: “G8 Petition with End Water Coalition” - Policy makers/Petition/UK – World
46. 2008 University of Berkeley: “I LOVE TAP WATER” Public/Visual – Stickers/USA – California
47. 2008 City of Toronto: “We’re drowning in platics” – Public/Visual in the city/Canada – Toronto
48. 2008 Tappening: “Every year, 38 billion bottles end up buried in the earth” – Public/Visual/ USA
49. 2008 WASH Africa: “WASH news Africa” – Policy makers – Water Professionals/Internet/ Africa – Extensive list of initiatives and news including campaigns in Africa - http:// washafrica.wordpress.com/
50. 2008 WSSCC: “Hurry up” – Policy makers – Public/Visuals/World
51. 2008 ONEP: “Water is life”/“Follow Miha” – Kids/Leaflet/Morocco
52. 2008 Brita: “Filter For Good” – Public – Customers/Visual and Internet/USA – World
53. 2008 Water Basins/Ministry of Ecology: “Water Is life” – Public/Consultation/France
54. 2008 WWF: “Polluted water kills as many people as nuclear explosion” – Public – Donors/ Visual/USA - World
55. 2008 Expo Zaragoza 2008: “Water and Sustainable Development” – Public/Event – Exhibition/ Spain - World
56. 2008 Grupo RPP – JICA – WSP: “Take care of water” – Public – Urban cities/Radio – TV programs/Peru
57. 2008 Tappening: “Obama and Mc Cain have a drinking problem” – Policy makers/Visual/ USA
58. 2008 Tap’D Ny: “New York Tap Water” – Public/Product/USA – New York
59. 2008 The Standard: “Water on Tap” – Public – Café and Restaurants/Label/London - UK
60. 2008 Helvetas: “Toilet Day: Poovertising” – Public/Street event/Lausanne - Switzerland
61. 2008 Scottish Water: “It costs you less than £1 a day. Enjoy” – Public/Visual Scotland - UK
62. 2009 World Water Day: “Transboundary Waters” – Mutli-public/Multi-supports/World
63. 2009 Solidarités International: “Water that saves, Water that kills” – Journalists/Press release/ France
64. 2009 Cristalline: “Water quiz” – Public – Customers/Internet/France
65. 2009 Union for mineral water. “Natural Mineral Water” – Public – Customers/Visual/France
66. 2009 Sanitation for Dijon: “There is not only Great Bourgogne in carafe” – Public/Visual/ France
67. 2009 UNICEF Tap Water Project and ebay ‘GivingWorks” – Public – Donors/Internet/World/ USA
68. 2009 UNICEF Tap Water Project: “Desperate” – Public – Donors/TV spot/USA
69. 2009 UNICEF Tap Water Project: “Dirty water” – Public – Donors/Events – Press relations/ USA
70. 2009 Charity Water: “The September Campaign” - Public – Donors/TV spot/USA

71. 2009 GreenPeace: "You're driving coastlines under water"/Visual/World
72. 2009 Tappening: "If bottled water companies can lie, we can too" – Public/Visual/USA
73. 2009 Evian: "Baby Ballet on Roller" - Public – Customers/Tv spot/France - World
74. 2009 Association of California Water Agencies: "Save our water" - Public/TV – Radio – Social media/Public/California - USA
75. 2009 NFB: "Waterlife" – Public/Internet – Documentary/Canada - http://waterlife.nfb.ca/
76. 2009 EPA: "Water on Tap, What you need to know" – Public/Brochure in English and Spanish/Chinese/Europe
77. 2009 Green Thing: "Tears of a Cloud" – Public/TV Spot/World - http://vimeo.com/4546290
78. 2009 WWF: "Waterprinting"/Public – Donors/Visual/World
79. 2009 Join the pipe: "Join the pipe community" – Public – Donors/Internet – TV spot/The Netherlands - World
80. 2010 World Water Day: "Clean water for a healthy" - World/Mutli-public/Multi-supports/World
81. 2010 UN Water: "Water Smart" – Public/Visual/World
82. 2010 WaterAid: "World's Longest Toilet Queue" – Public – Policy makers – Donors/Event: people queuing/UK
83. 2010 Solidarités International: "Water wall" – Public/Event: Wall in the city/France – Paris
84. 2010 Lyonnaise des Eaux: "New ideas for Water" – Public – Policy makers/Internet - Book/France
85. 2010 Charity Water: "What happens next" – Public/Internet/USA
86. 2010 Charity Water: "Unshaken" – Public – Donors/TV spot/USA/Haiti
87. 2010 Quicksilver and Roxy: "No water sucks" – Public/Visual and contest/World
88. 2010 Fondation France Libertés: "Water Leaf" and "Water, a human good" – Public – Donors/Visual/petition/France
89. 2010 Waterwise, Thames Water, WWF: "Save Water Swindon"/Public/Internet/UK Swindon, www.savewaterswindon.org.uk
90. 2010 Green Thing: "World Tap Water Week" – Public/Internet/World
91. 2010 Helvetas: "What to do when water is nocive?" Public/TV spot/Switzerland and "Baby bottle and dirty water" – Policy makers/Event: 4,000 baby bottles in the streets/Zurich - Switzerland
92. 2010 Water makes money – Documentary/Europe/World
93. 2010 WSSCC: "Listen, share, act!" – Public/Radio/Countries with low literacy rates - World
94. 2010 The story of stuff project: "The Story of Bottled Water" - Public/TV spot – Internet/World - www.storyofstuff.org
95. 2010 Bottle Water Matters: "The Real Story of Bottled Water – Public/TV spot – Internet/World - www.bottledwatermatters.org
96. 2010 WWF: "Water Futures" – Public/Visual/World
97. 2010 Okanagan WaterWise: "Make water work" – Public – Residents – Kids/Internet – Visual/Okanagan Valley - Canada
98. 2010 WaterAid: "Dirty Leaders, Dirty Water" – Public/TV spot/World
99. 2010 Expo 2010 Shanghai: "Hao Bai" – Public – Policy makers/Logo, exhbitions, events/China - World
100. 2011 World Water Day: "Water for cities: responding to the Urban Challenge" Mutli-public/Multi-supports/World

101. 2011 Duravit: "Water is Life" - Public – 5,000 artists/Exhibition – Visual/Nanjing Water Day – Cape Town/83 countries - World
102. 2011 UNICEF Tap Project: "Celebrity Water" – Public/Internet - TV spot – Bottles/USA
103. 2011 UNICEF and Acqua for Life Giorgio Armani: "Acqua for Life" – Public – Donors/Internet/World/USA/Europe - www.acquaforlife.org
104. 2011 Syndicat pour la promotion des communes de Polynésie Française: "Topota" – Public – Policy makers/Information – Panel/Polynesia Islands
105. 2011 Solidarités International: Water Ink – Public – Donors/TV spot/France – World
106. 2011 Helvetas: "Water for food" – Public – Donors/Tv spot/Switzerland
107. 2011 Lyonnaise des Eaux "Contract for the Health of Water" – Policy makers –Public/Brochure/France
108. 2011 WaterAid/WSSCC: "Walk for Water" - 300,000 people/Petition/UK – World
109. 2011 ONEMA: "Change our point of view on water" – Public/Internet/France
110. 2011 SEDIF: "More than words, engagement for water" – Customers/Visual – Internet – Mail/Paris - France
111. 2011 EPA: "Wasting Water is Weird" - Public/TV spot – Internet/USA - World
112. 2011 GiveMeTap: "Give Me Tap"/Public/Internet, restaurant/UK cities
113. 2011 SUEZ Environnement: "Making water safe to drink is good business' – Visual -Internet/France
114. 2011 Veolia Environnement: "Growing Blue" - Policy makers – Business/Internet/World - http://growingblue.com/the-growing-blue-tool/
115. 2011 Volkswagen: "Think Blue" - Customers/Visual – Internet - Spots/World
116. 2011 FLOW: "I am for Love of Water" – Public/Internet – Tv spot – Event/South Africa
117. 2011 Yorkshire Water: "CIPR Pride Awards" – Public – Customers/Visual – TV – Web/UK – Yorkshire
118. 2011 European Commission: "Generation Awake" – Public/Website/Europe - www.generationawake.eu/en
119. 2011 Disputed Waters: "Disputed Waters" - Public/Web-Documentary/The Netherlands – France – World - www.disputedwaters.com
120. 2011 Renew: "Save water, energy, money" – Public/TV spot/EU funded Project/UK – Europe
121. 2011 European Water Patnership: "European Water Stewardship" - Industries – Policy makers/Events – Label/Europe
122. 2011 Australian Government: "Water for the future campaign" – Public – Farmers - Communities/Press – PR – TV – Radio – Outdoor/Murray-Darling Basin - Australia - www.environment.gov.au/water
123. 2012 World Water Day: "Water and Food" - Mutli-public/Multi-supports/World
124. 2012 UNICEF Tap Water Project: "Make a reservation to save the world"/Public/USA
125. 2012 European Commission: "The Blueprint to safeguard Europe's Water" – Policy makers – Basins agencies – governments/Press Release – Event/Europe
126. 2012 SIG – UNICEF: "Tap water for the Geneva's Marathon" – Public/Tap water distribution – Press release/Geneva - Switzerland
127. 2012 State of Colorado: "H2O Jo: the Keep it clean" campaign – Public – Residents/Materials – Events/Colorado - USA - http://npscolorado.com/h2ojo.htm
128. 2012 UNEP: "Drop by Drop" - European public/Internet/Europe – World - www.dropbydrop.eu
129. 2012 Environment Canada (Government): "Great Lakes Highlights" – Public – Policy makers/Internet – Events – TV News/Canada

130. 2012 EEA: "If water could speak" - Public/TV spot/Europe - World
131. 2012 Solidarités International: "The Minute for Water"/Public – Policy Makers/Social media/ France – World - www.solidarites.org
132. 2012 Thames Water: "We are in Drought" – Waterwisely – Public/Visual – Internet/UK - http://secure.thameswater.co.uk/drought
133. 2012 Angela Morelli: "Virtual Water" - Public/Website/World - www.angelamorelli.com/ water
134. 2012 Every Last Drop - Public/Website/UK – World - http://everylastdrop.co.uk/
135. 2012 Eau de Paris: "Under the water" - Public/Webdocumentary/France - http://webdoc. france24.com/bains-douches/
136. 2012 European Public Services Union: "Right for water" – Public – Policy makers/Petition – Internet – TV spot/Europe: www.right2water.eu
137. 2012 French Embassy in the United States: "International Film Project 2012" – Children – Policy makers/Movies/France – India – USA – Brazil: http://sciencefestival2012. blogspot.fr/
138. 2012 SUEZ Environnement: "Water, a challenge for the 21st century" – Public – Water professionals/Iconography/World
139. National Association of Water Companies: "Water is your business" – Public/Infography – Internet/USA http://waterisyourbusiness.org/
140. Water for Jobs: "Water puts America to work" – Policy makers/Internet – Video – Events/ USA - www.waterforjobs.org
141. 2012 WaterAid: "1 in 3 women" - Public – Donors/TV Spot/World
142. 2012 WaterAid: "Dirty leaders, Dirty water" - Policy makers – G8/Tv spot/World
143. 2012 Barakacity: "L'eau c'est la vie" - Water is life - Public – Muslim communities/Internet fundraising with TV spot/France – Togo
144. 2012 WWF: "Water saving products" – Public – Donors/Product/China – USA -World
145. 2012 WWF – H&M: "New Water Strategy" - Public – Donors/Visual/World
146. 2012 WWF: "Water Risk Filter" - Public – Donors/TV spot - Internet/World
147. 2012 Governement of Ethiopia: "Grand Millennium Dam" – Public/Website - Ethiopia – Nile Basin – World
148. 2012 Whole Living (Magazine): "Test your Water IQ" – Public/Iconography/USA
149. 2012 Conseil Régional de Seine et Marne: "Water is not a luxury" – Public/Visual/France
150. 2012 Evian: "Live Young" – Public – Customers/Visual/World
151. 2012 UTEC: "Water Panel" – Public – Residents/Advertising panel/Peru
152. 2012 Syndicat Intercommunal du Centre et du Sud de la Martinique: "The secret of my beauty? Tap Water" – Public/Visual/Martinique islands
153. 2012 Democratie & Communication: "A vote can change the result" – Public/Visual/France
154. 2012 Association des Maires de France: "Francophone women and water" – Policy makers/ Radio documentary/France – West Africa
155. 2013 World Water Day: International Year of Cooperation – Various target/World
156. 2013 Solidarités International: "Water ink for your water drop" – Public – Policy makers – International audience/TV spot – Internet/World
157. 2013 Water.org: "I'm going on strike" – Public – Donors/Matt Damon, TV press conference – Internet/World
158. 2013 UNICEF: Tap Water Project: "Water network" – Public/Social Media/World
159. 2013 Evian: "Baby and Me" – Public – Customers/TV spot/World

160.	2013	Waterwise: "Water Efficiency Initiatives" – Policy makers – Utilitlies/Brochure/UK
161.	2013	AFD: "Water, a source of future for Yahimbaé" – Public - Policy makers/Video – Internet/France
162.	2013	Lifebuoy: "My son has turned five today" – Public/TV Spot/India - World
163.	2013	Saatchi & Saatchi: "Farmageddon" – Public – UN Water – Policy Makers/Game/World
164.	2013	Whole World Water: "Call for Whole World Water by Richard Branson" – Public – Donors – Restaurants & Hotels Industries/Internet – Video/World
165.	2013	SIG: "Stupefying!" – Public/TV spot/Geneva - Switzerland

IWA Publishing's authorised EU representative for General Product Safety Regulations is Diane D'Arras, 15 rue Duret, 75116 Paris, France, e-mail: safety@iwap.co.uk.

Printed and bound by CPI Group (UK) Ltd, Croydon, CR0 4YY
16/06/2026
02140400-0002